SELECTIVE PLASMA COMPONENT REMOVAL

edited by

ALVARO A. PINEDA, M.D.

Director, Apheresis Laboratory
Mayo Clinic Blood Bank and Transfusion Service
Associate Professor of Laboratory Medicine
Mayo Medical School
Consultant, Department of Laboratory Medicine
Mayo Clinic and Mayo Foundation
Rochester, Minnesota

 Futura Publishing Company
Mount Kisco, New York
1984

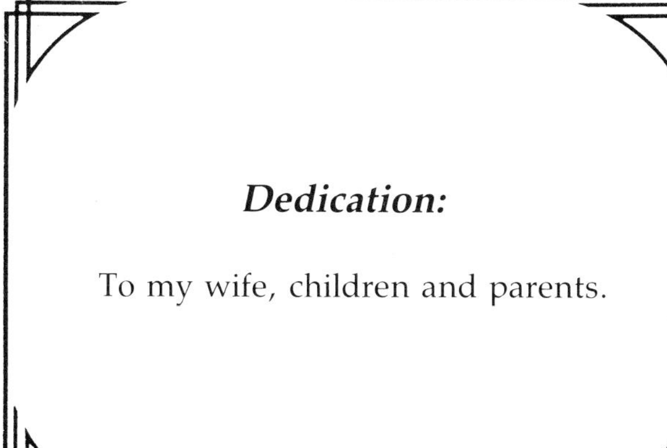

Dedication:

To my wife, children and parents.

Contributors

William I. Bensinger, M.D.   Assistant Professor of Medicine, Fred Hutchinson Cancer Research Center and the University of Washington School of Medicine, Seattle, Washington

David H. Bing, Ph.D.   Senior Investigator, Center for Blood Research, Boston, Massachusetts

Christoph Bode, M.D.   Post-doctoral Fellow, University of Cologne, Cologne, West Germany

Charles A. Bowles, Ph.D.   Director, Department of Experimental Therapeutics, ImmuQuest Laboratories Corp., Rockville, Maryland

Edwin A. Burgstaler, M.T. (ASCP)   Development Technologist, Apheresis Laboratory, Mayo Clinic Blood Bank, Department of Laboratory Medicine, Mayo Clinic and Mayo Foundation, Rochester, Minnesota

E. Rolland Dickson, M.D.   Professor of Medicine, Department of Internal Medicine, Division of Gastroenterology and Internal Medicine, Mayo Clinic and Mayo Foundation, Rochester, Minnesota

Jacqueline Foidart, M.D., Ph.D.   Consultant, Assistant, Nephrology Unit, Universite de Liège, Hôpital de Baviere, Liège, Belgium

Francis Frankenne, Ph.D.   Assistant, Head Chemist, Endocrine Unit, Université de Liège, Hôpital de Bavière, Liège, Belgium

Georges Hennen, M.D., Ph.D.   Consultant, Assistant, Nephrology Unit, Université de Liège, Hôpital de Bavière, Liège, Belgium

Takashi Horiuchi, M.S.   Research Fellow, Department of Artificial Organs, Cleveland Clinic Foundation, Cleveland, Ohio

Jean-Pierre Kinet, M.D.   Assistant, Department of Medicine, University of Liège, Liège, Belgium

Philippe Mahieu, M.D., Ph.D.   Assistant Professor, Chief, Nephrology Unit, University of Liège, Hôpital de Bavière, Liège, Belgium

Paul S. Malchesky, M.S.   Director of Metabolic Assistant Program, Department of Artificial Organs, Cleveland Clinic Foundation, Cleveland, Ohio

Gerald Messerschmidt, M.D.   Co-Director, Immunobiology Laboratory, Department of Medicine, USAF Medical Center, Lackland Air Force Base, Lackland, Texas

Yukihiko Nosé, M.D., Ph.D.   Chairman of Department, Department of Artificial Organs, Cleveland Clinic Foundation, Cleveland, Ohio

Alvaro A. Pineda, M.D.   Director, Apheresis Laboratory, Mayo Clinic Blood Bank and Transfusion Service, Associate Professor of Laboratory Medicine, Mayo Medical School, Consultant, Department of Laboratory Medicine, Mayo Clinic and Mayo Foundation, Rochester, Minnesota

Gottfried Schmer, M.D.   Professor of Laboratory Medicine and Biochemistry, Department of Laboratory Medicine, University of Washington, Seattle, Washington

James W. Smith, M.D., Ph.D.   Director of Clinical Apheresis Program, Department of Artificial Organs, Cleveland Clinic Foundation, Cleveland, Ohio

Wilheim Stoffel, M.D., Ph.D.   Director of the Institute of Physiological Chemistry, University of Cologne, Cologne, West Germany

Howard F. Taswell, M.D.   Professor of Laboratory Medicine, Department of Laboratory Medicine, Mayo Clinic and Mayo Foundation, Rochester, Minnesota

Introduction

This book responds to a perceived need for a publication that brings together, in a single volume, a description of the current status of a new field of scientific interest: selective removal of plasma constituents. Out of necessity, the chapters describe topics at varying stages of development, which makes uniformity difficult to achieve. The book reflects a cross-section of the prevailing thinking relative to the selective extraction of plasma components. Although fractionation of plasma has been attempted in the past, the impetus for the removal of isolated constituents was provided only recently by an increase in the use of plasma exchange—therapeutic plasmapheresis. Plasma exchange, the wholesale removal of plasma as a means of removing specific plasma constituents for the purpose of modifying disease, is unselective and wasteful.

This book attempts to report in great detail methods that permit the selective removal of various plasma constituents that have been assigned a putative role in a disease process. Description of the methods for extracting bile acids, low density lipoproteins (LDL), IgG and immune complexes, amino acids, macromolecules, and A and B isoagglutinins form the core of the following pages. The list of constituents is short in relation to the large number of plasma substances known to exist; however, the potential applications of even such a short list are of great significance. The substances removed and the responses elicited by their extraction have important medical and financial implications.

These implications are vast whether one focuses on the treatment of malignant diseases by amino acid deprivation or by protein A binding or on the treatment of xanthomas and arteriosclerosis of the hypercholesterolemic patient or on the treatment of autoimmune disease by macromolecule removal and cryogelation. Also noteworthy is the significant modification of the plasma milieu resulting from the selective removal of AB isoagglutinins in preparation for transplantation of an ABO-incompatible bone marrow. The selective removal of target plasma substances and the return of the patient's own modified plasma result in important economies.

The selective removal of LDL is explored in the opening chapter by Stoffel and Bode. They describe the culmination of intensive work revolving around the preparation of an affinity column for the specific removal of LDL. Animal studies had shown that after production and isolation from sheep blood of monospecific antibodies against swine LDL, an effective and reusable plasmaperfusion system could be developed. By linking sheep antihuman LDL antibodies covalently to Sepharose, a system was assembled to treat hypercholesterolemic humans. Removal of up to 10 g of LDL per treatment with no side effects is reported. More importantly, no sheep LDL antibodies have been detected in the humans treated so far.

An affinity column was developed to treat the intractable pruritus of cholestasis. The column contains activated charcoal-coated glass beads that have greater affinity for bile acids and bilirubin than do other charcoal preparations, resins, and albumin. The successful treatment of patients with severe cholestasis and intractable pruritus unresponsive to phenobarbital and cholestyramine is described in the second chapter. The efficacy of the plasmaperfusion treatment is compared to that of dialysis and plasma exchange. Use of this column has resulted in significant amelioration of pruritus, thus creating a safe and reasonably effective modality of treatment in the absence of a therapeutic alternative.

The chapter by Bensinger gives a detailed account of work resulting in selective removal of anti-A and anti-B isoagglutinins, whether they are IgG or IgM, by an affinity column that uses synthetic sugars linked to crystalline silica as the sorbent. Considerable reduction in isoagglutinin titers is obtained, which circumvents the need for intensive plasma exchange in patients destined to receive an ABO-incompatible bone marrow transplant. Removal of the antibodies eliminates the risk of hemolysis of the large number of erythrocytes that contaminate the bone marrow preparation. The column can be desorbed efficiently to yield high antibody concentrations with potential commercial value. This is one of several immuno-adsorbent systems described in the book for the extraction of circulating immune reactants.

The recognition that depletion of a number of amino acids from human plasma in vivo has therapeutic implications led to the development of techniques, aptly described by Schmer, for selective amino acid removal. The applications vary from the well-defined effects of removing L-asparagine, L-glutamine, and L-tryptophan in cancer and the removal of L-asparagine to achieve immunosuppression to the less well-defined effect of removing tyrosine from patients in hepatic coma. Although a therapeutic effect has been obtained by injecting the corresponding depleting enzyme, undesirable toxic and immunologic responses have limited the application. Thus, an alternative method capable of having a therapeutic effect yet devoid of toxicity and adverse immunologic response was needed. Whether amino acid deprivation becomes an established form of therapy in oncology, hepatology, and renal transplantation remains to be seen.

The protein A produced by some strains of *Staphyloccoccus aureus* has a strong affinity for the Fc portion of IgG-1, -2, and -4. This property has been exploited in studies aimed at producing affinity columns or filters capable of removing IgG and immune complexes selectively from plasma. Plasmaperfusion of these affinity columns of filters has been applied to the treatment of human and animal malignancies as described in Chapters 5 and 6. Both chapters report a tumoricidal effect for which the mechanism is not exactly known although a number of hypotheses have been advanced. Removal of a tumor-blocking activity and immune complexes from serum are often cited hypotheses to explain the tumoricidal effect but, as pointed out by the authors, the small amounts of protein A used would not remove significant amounts of immune complexes. Also explained are the tumoricidal effect of intravenously injected protein A and the induction of interferon production and activation of complement and B and T cells that follow treatment with protein A.

The chemical and physical precipitation of plasma constituents, mainly immune globulins, are dealt with in the last two chapters. Bing describes treatment of plasma with zinc diglycinate to cause rapid precipitation of immune globulins and immune complexes which can readily be excluded by centrifugation. The precipitate is devoid of albumin, haptoglobin, complement, and other useful plasma constituents. Horiuchi et al. describes a less selective but efficient approach—cryogelation with filtration of macromolecules. This system excludes any molecule larger than 100,000 daltons while allowing passage of albumin which is used for autologous oncotic replacement. Unlike Bing's method, which at this point is off-line, removal of macromolecules is achieved on-line from plasma that has been obtained from whole blood by filtration in a membrane separator with high permeability.

It will become obvious to the reader that the techniques described in this book are still in an experimental phase and consequently have limitations that need to be overcome before they will gain wide acceptance. Questions remain regarding filter or column effluents, modified plasma composition, and the short-term and long-term effects on the recipient. Equally important, the reactions encountered in the course of perfusion require thorough investigation. The specific removal of plasma constituents is likely to be the hemapheresis of the future. In theory, there should be a method for the specific removal of each of the numerous plasma constituents. The limiting factor lies not in the technology of removal but rather in the identification of the unwanted plasma component that is responsible for the pathologic process.

ALVARO A. PINEDA, M.D.
Mayo Clinic
Rochester, Minnesota

Contents

CHAPTER 1

Selective Removal of Low Density Lipoproteins

Wilheim Stoffel, M.D., Ph.D.
Christoph Bode, M.D.

INTRODUCTION

Lipoproteins and Lipid Transport

Nonpolar lipids, such as triglycerides, cholesterol, and cholesteryl esters, are transported in the plasma as lipoproteins. In plasma, these hydrophobic molecules are associated with apolipoproteins and are packed in the core of the particle. A surface coat of polar lipids, such as phospholipids and unesterified cholesterol, together with specific apoproteins, surrounds this hydrophobic core. In addition to the binding of lipids, apoproteins interact with a number of specific enzymes or receptors on cell membranes.

Apoproteins represent the antigenic components of the lipoproteins. Table 1 gives the characteristics of the five major lipoprotein classes that normally circulate in human plasma. These lipoprotein classes differ in the composition of nonpolar lipids in the core and in the composition of polar lipids and apoproteins. Consequently, physical properties such as density, size, and electrophoretic mobility are different.

Table 1
Characteristics of the Major Classes of Lipoproteins in Human Plasma

Lipoprotein class*	Major core lipids	Major apo-proteins	Density, g/ml	Diameter, Å	Electrophoretic mobility
Chylomicrons	Dietary triglycerides	AI, AII, B, CI, CII, CIII	<1.006	800−5,000	Remains at origin
VLDL	Endogenous triglycerides	B, CI, CII, CIII, E	<1.006	300−800	Pre-β
Remnants	Cholesteryl esters, triglycerides	B, CIII, E	<1.019	250−350	Slow pre-β
LDL	Cholesteryl esters	B	1.019-1.063	180−280	β
HDL	Cholesteryl esters	AI, AII	1.063-1.210	50−120	α

*VLDL = very low density lipoprotein; LDL = low density lipoprotein; HDL = high density lipoprotein.

Lipoproteins and Atherosclerosis

In most western countries cardiovascular diseases have become the main cause of death.[1] Atherosclerosis is the underlying process of most deaths related to myocardial infarction and cerebrovascular diseases. From extensive epidemiologic studies, three major risk factors have emerged: increased plasma chlolesterol concentration, hypertension, and cigarette smoking.[2]

In normal subjects the distribution of total serum cholesterol is approximately 70% in low density lipoproteins (LDL cholesterol) and 25% in high density lipoproteins (HDL cholesterol). There is a direct correlation between the increase in plasma LDL and the incidence of atherosclerosis; this correlation is inverse for plasma HDL.[3]

Several dyslipoproteinemias have been identified, supporting the view that lipoproteins containing apolipoprotein B (apo B) constitute an independent risk factor for cardiovascular disease, whereas a protective effect is attributed to HDL. Homozygous subjects who have familial hypercholesterolemia rarely survive past the early part of their second decade,[4] whereas subjects with hypobetalipoproteinemia or hyperalphaproteinemia have a life span considerably longer than that of normal individuals.[5,6]

A therapeutic method leading to serum depletion of apo B-containing lipoproteins but with retention of α-lipoproteins would therefore be a most desirable tool to impede the development or progression of atherosclerotic lesions.

FAMILIAL HYPERCHOLESTEROLEMIA

Genetic Aspects

Familial hypercholesterolemia is a common autosomal dominant disorder that affects approximately one in 500 persons in the general population. It is associated with a type IIa lipoprotein pattern phenotype.[7]

The primary defect resides in the gene for the LDL receptor. Three types of mutant alleles have been described in this locus.[8] The most common, labeled $R\,bO$, specifies a nonfunctional gene product, which means that the LDL receptor exhibits a complete defect. The second most common, named receptor defective ($R\,b-$), produces a receptor that has 1–10% of normal LDL-binding activity. The third type, $R\,b+;iO$, produces a receptor that binds LDL normally but is unable to internalize the LDL receptor complex.

Homozygotes have two mutant alleles at the LDL receptor locus and hence their cells are unable to bind or take up LDL. Heterozygotes possess one functioning and one mutant allele, and therefore their cells are able to bind or take up LDL but at about half the normal rate.

Biochemical Aspects

Normal cells produce the surface LDL receptor that is responsible for the uptake of LDL cholesterol from plasma. This cholesterol is utilized for the synthesis of cell membranes in any cell, of bile acids in the hepatocyte, or of steroid hormones in the adrenal cortex and ovary. In heterozygotes and homozygotes, the reduced number of receptor molecules is compensated for by an increase in the plasma LDL level; in the steady state, a normal amount of LDL is degraded but at the price of a manyfold elevation in plasma LDL levels.[9] Thus, familial hypercholesterolemia is due to a defect in the receptor that mediates removal of LDL from plasma. Besides the impaired catabolism, a two- to threefold overproduction of LDL has been noted in homozygotes.[10-12] The underlying mechanism is the lack of a negative feedback control of the key enzyme in cholesterol synthesis, β-hydroxymethylglutaryl-CoA reductase. High LDL concentrations in plasma lead to an enhanced uptake of LDL by scavenger cells (macrophages, smooth muscle cells) which accumulate at various sites in the body, including the arterial wall, and produce the clinical symptoms—namely, atherosclerosis.

Clinical Aspects

The heterozygote form of the disease is characterized clinically by a two- to threefold increase in plasma LDL cholesterol, present at birth and persisting throughout life.

Xanthomas of the tendons, especially the Achilles tendon and the extensor tendons of the hand, develop in about 80% of heterozygotes. They represent the deposition of LDL-derived cholesteryl esters in tissue macrophages. Less specific but equally easy to diagnose are cutaneous xanthomata, xanthelasma, and arcus lipoides corneae; these are present in about 50% of heterozygous patients. The most important clinical feature, however, is the occurrence of premature coronary atherosclerosis. Of affected men, 85% have experienced a myocardial infarction by age 60, with a peak incidence in the fourth and fifth decades. In women, the manifestation of the disease tends to be delayed by 10 years. It has been estimated that familial hypercholesterolemia is responsible for about 5% of all myocardial infarctions.[10]

In homozygous patients, LDL cholesterol is increased six- to eightfold. Besides the symptoms described above, these patients display characteristic yellow plaque-like cutaneous xanthomata with predilection sites at the interdigital skin, knees, elbows, and buttocks. Coronary heart disease usually develops before age 10. Homozygotes rarely survive their second decade.[13]

Established Treatments for Heterozygotes

The goal of therapy in familial hypercholesterolemia is to reduce the concentration of plasma LDL while maintaining a sufficient supply of cholesterol to the cells. The ideal approach is to increase the number of LDL receptors. The possibility of enhancing the production of LDL receptors by the cell does exist in heterozygotes, by stimulating the single normal receptor gene. This can be achieved by feeding the patient a diet low in cholesterol and saturated fats and high in unsaturated fats. On a strict diet, plasma LDL can be reduced by about 15% in heterozygotes.[14]

Cholestyramine and Cholestipol, bile acid-binding resins, trap bile acids excreted by the liver and thus interfere with the enterohepatic circulation. As a result, the liver converts more cholesterol to bile acids, which depletes the hepatic cholesterol content. In order to maintain a sufficient cholesterol content for the activated metabolic pathways, the liver produces more LDL receptors, thus lowering the plasma LDL level.[15] However, hydroxymethylglutaryl-CoA reductase is activated,[16] which leads to an increased synthesis of cholesterol, thereby partly offsetting the need for more LDL receptors. On the whole, the therapeutic result is minimal. A reduction of plasma LDL by about 25% is possible in heterozygotes.

Nicotinic acid has been used to suppress the de novo synthesis of cholesterol in the liver. New, considerably more potent drugs will be available in the future—namely, compactin and mevinolin.[17-19] Given together with Cholestipol, mevinolin is able to decrease the plasma LDL level by about 75%.[15]

The above drug regimens are accompanied by side effects such as bloating, cramps, and constipation with the bile acid-binding resins and hepatotoxicity with nicotinic acid. Compactin and mevinolin are under clinical trial at present.

Surgical methods have been proposed to lower plasma LDL in heterozygotes. Creation of an ileal bypass prevents the reabsorption of bile acids, as cholestyramine does, but it has proved to be no more effective than the drug.[20]

All of the above procedures are ineffective in homozygous hypercholesterolemia because there is no functioning LDL receptor gene to be stimulated.

METHODOLOGICAL BACKGROUND OF PLASMA LDL APHERESIS BY LDL IMMUNOADSORPTION

Essentials

The requirements that a procedure has to meet in order to be effective during lifelong therapy are the following:

(1) The procedure for the removal of the pathogenic factor must be highly specific.
(2) It must be capable of continuously depleting the plasma of the component under study.
(3) It must be rapid and efficient.
(4) The physical status of the patient under treatment should remain unimpaired. The procedure should be as minimally invasive as possible—e.g., no arterio-venous shunt operation, no arterial puncture.
(5) It should be amenable to efficient use to make the system financially feasible.

Preparation of LDL

LDL is isolated by ultracentrifugation.[21] The density range 1.006–1.063 g/ml is screened for homogeneity by agarose gel electrophoresis,[22] immunodiffusion,[23] immunoelectrophoresis,[24] and electron microscopy.

Preparation of LDL-Sepharose Cl-4B

This procedure has been published.[25] Sepharose Cl-4B (30 ml) is washed with water and suspended in 90 ml of ice-cold 1 M sodium carbonate. Cyanogen bromide (6 g in 3 ml of acetonitrile) is added with stirring and the activated Sepharose beads are collected on a fritted glass funnel after the cyanogen bromide has dissolved completely (2–5 minutes). The Sepharose is washed first with 500 ml of ice-cold 0.2 M bicarbonate buffer at pH 9.5 and then with 500 ml of ice-cold 0.5 M bicarbonate buffer at pH 8.5 and immediately resuspended in a solution of 300 mg of LDL (60 mg of apo B) in 60 ml of 0.5 M bicarbonate buffer at pH 8.5. The mixture is incubated for 20 hours at 4°C. The LDL Sepharose is collected on a glass funnel and suspended in 60 ml of phosphate-buffered saline/2 M glycine and stirred for 12 hours at 4°C; during this period, active sites on the Sepharose are deactivated. Finally, the LDL Sepharose beads are washed with 2 liters of phosphate-buffered saline. Approximately 7.5 mg of LDL (1.5 mg of apo B) per ml of Sepharose is covalently linked.

The activation reaction of Sepharose with cyanogen bromide produces a reactive imidocarbonate on the Sepharose:

$$\begin{array}{c} | \\ -OH \\ \\ \quad + \text{ CNBr} \longrightarrow \\ \\ -OH \\ | \end{array} \qquad \begin{array}{c} | \\ -O \\ \\ \quad C = NH \\ \\ -O \\ | \end{array}$$

The activated Sepharose will then couple to an amino group on the protein to yield the immobilized product. Coupling of the protein is mainly through the ε-amino group of lysine and the α-amino group of the amino-terminal amino acid. The multipoint attachment between protein and matrix increases the stability.[26]

Isolation of Sheep Anti-LDL

Monospecific anti-LDL antiserum is mass produced in sheep immunized with human serum LDL. The antibody is isolated by a single chromatographic step: 50 ml of heparin-treated sheep plasma is diluted with 50 ml of isotonic saline and recycled for 4 hours over an LDL Sepharose Cl-4B column (bed volume, 250 ml). Unbound proteins are eluted with saline, and adsorbed antibody is eluted as a single band with 0.2 M glycine/HCl buffer at pH 2.8.

Preparation of Antihuman LDL Sepharose Cl-4B

Antihuman LDL (5 g in 500 ml of 0.2 M bicarbonate buffer at pH 9.5) is added to 500 ml of Sepharose Cl-4B activated by treatment with 100 g of cyanogen bromide in 50 ml of acetonitrile. The remainder of the procedure is as described for preparation of LDL Sepharose.

For the treatment of patients, silanized glass columns (internal diameter, 10 cm; length, 8 cm) with sintered glass filters (pore size, 20–40 μm) are packed with 400 ml of anti-LDL Sepharose. Each set of two columns is assigned to a single patient and used only for the treatment of that patient.

Treatment of Patients

Additional details have been published elsewhere.[27,28] Preliminary experiments in animals required a shunt operation between the carotid artery and the jugular vein to obtain an effective blood flow. In humans, vascular access is secured via the antecubital veins of the arms. The patient's blood enters the continuous-flow blood cell separator at a flow rate of 60–80 ml/min. Anticoagulation is achieved with heparin (2,500 units administered before the run and then 40 units/min up to a total of 10,000 units per treatment) and citrate (added to the blood at a ratio of 1:18). The plasma is passed through the anti-LDL Sepharose bed at a flow rate of 30–40 ml/min and then, free from LDL, is recombined with the blood cells to enter the antecubital vein of the opposite arm. The flow to the two columns is alternated. While LDL is being adsorbed by one column, the other column is being regenerated outside the circuit. Regeneration is achieved by desorption with 0.2 M glycine/HCl buffer at pH 2.8, neutralization with phosphate-buffered saline, and equilibration with 0.9% NaCl.

When a freshly regenerated column is being loaded with plasma, the saline in the dead volume (about 40 ml) is directed into a waste container. Before a column is switched from "absorption" to "regeneration," the plasma in the dead volume of the LDL-loaded column is pushed into the patient with 400 ml of saline.

Even when exercising great care, dilution of the patient's plasma by about 10% cannot be avoided. This dilution adds to the 10% dilution caused by the cell separation procedure and the anticoagulants. In addition, withdrawal of blood samples from the patient leads to a loss of plasma and blood cells. In animal experiments, a reduction of plasma cholesterol of 15.9% and of plasma total protein of 17.5% was observed when the treatment procedure was performed with Sepharose columns not carrying bound antibodies. These dilutional effects have to be taken into account when the effectiveness of the procedure is being judged.

The working capacity of each column is about 2 g of total LDL cholesterol or about 1 g of apo B or 4 g of LDL. The length of time a column can be used before becoming saturated can be calculated from the plasma LDL level in the patient. It turned out to be reasonable to change columns after 30, 45, and 60 minutes. Between treatments the columns have to be filled with 0.02% sodium azide in phosphate-buffered saline saturated with chloroform and stored at 4°C. Before use, each column is rinsed with 5 liters of isotonic saline. Strict measures are taken to insure sterility. Before each treatment the saline eluate is tested for bacterial contamination and for pyrogens. Some columns have been used as many as 60 times over a period of 18 months without any loss of LDL binding activity.

RESULTS OBTAINED FROM CLINICAL STUDIES

A considerable decrease in plasma LDL cholesterol level is produced by LDL apheresis (Table 2). The treatment lowers the LDL level to less than that of normal individuals. The intervals between treatments varied from 1 to 5 weeks, but treatment seemed to be most efficient when carried out at 1-week intervals. Despite a rapid increase of plasma LDL after therapy, it is possible to establish a level well under 300 mg/dl if intervals do not exceed 10 days. On the average, 67% of the

Table 2
**Analysis of 139 Treatments in Six Patients: General Data
and Changes in Serum Lipids**

Feature	*Mean*
General information	
Period of treatment	8.5 months
Interval between treatments	9.7 days
Number of treatments	23 treatments
Number of loaded columns	3−5 columns
Duration of treatment	186 minutes
Amount of plasma treated	5,933 ml per treatment
Serum lipids	
Before start of therapy, total cholesterol	461 mg/dl; LDL-C, 411 mg/dl
Before a regular treatment, total cholesterol	347 mg/dl; LDL-C, 285 mg/dl
After a regular treatment, total cholesterol	135 mg/dl; LDL-C, 93 mg/dl
Between treatments	241 mg/dl; LDL-C, 189 mg/dl
Removed:*	
Total cholesterol	61.1%
LDL-C	67.3%
VLDL-C	54.7%
Triglycerides	53.6%
HDL-C	26.5%

*As percentage of original amount.

LDL cholesterol is removed from plasma by a single 3-hour treatment; 7−15 g of LDL is removed per session. Very low density lipoprotein (VLDL) cholesterol and serum triglycerides are also reduced by 50%. On the whole, this means that for a period of time right after treatment, mobilization of cholesterol from pools is likely to occur without a subsequent phase of extensive cholesterol deposition.

Figure 1 presents the time course of changes in the concentrations of serum lipids during a single immunoadsorption treatment. Total cholesterol decreased from 470 mg/dl to 112 mg/dl. The temporary changes in total protein concentra-

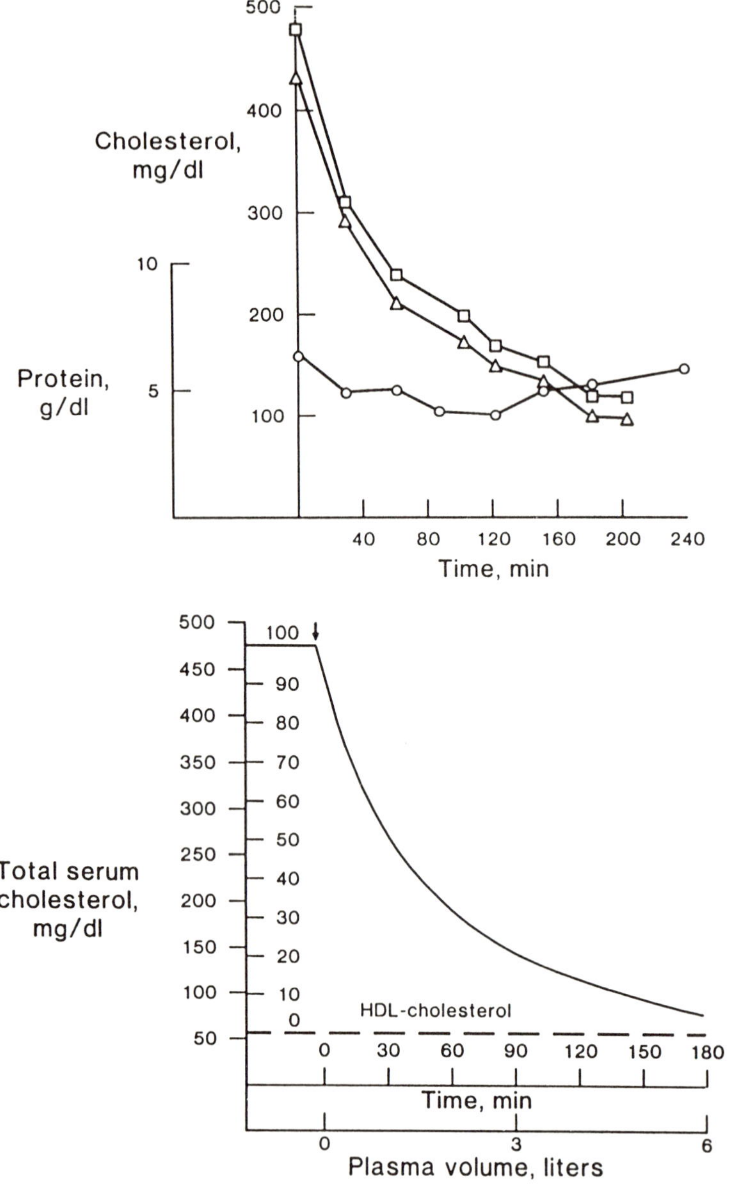

Figure 1: *Time course of changes.* Upper: *total cholesterol* (□), *LDL* + *VLDL* (△), *and plasma protein* (○) *in plasma line before entering immunoadsorption column.* Lower: *absolute concentration change and percentage removal of LDL cholesterol by immunoadsorbent column. Plasma flow rate: 30 ml/min.*

tions and in HDL cholesterol concentration were due to dilution. The actual loss of plasma components was minimal. Total protein concentration returned to the original level within 2–4 hours after treatment.

Figure 2 shows the changes seen in two successive LDL aphereses. After the first treatment, the total plasma cholesterol increased gradually to a new plateau in 8–10 days; then the second treatment was carried out.

Table 3 indicates that all plasma parameters did not change more than can be attributed to dilution. Only the leukocyte count represents an exception. The increase, mainly due to granulocytes, may be the result of a mild and temporary activation of the complement system.

Measurement of total acidic and neutral fecal steroid excretion revealed that, despite an impressive decrease in plasma LDL concentration, fecal steroid excretion remained unaltered (Fig. 3). Because lipoprotein synthesis did not change,[29,30] the constant fecal efflux and the depletion of plasma cholesterol suggest that cholesterol pools of the body are depleted.

After the first two treatments, a visible reduction of the

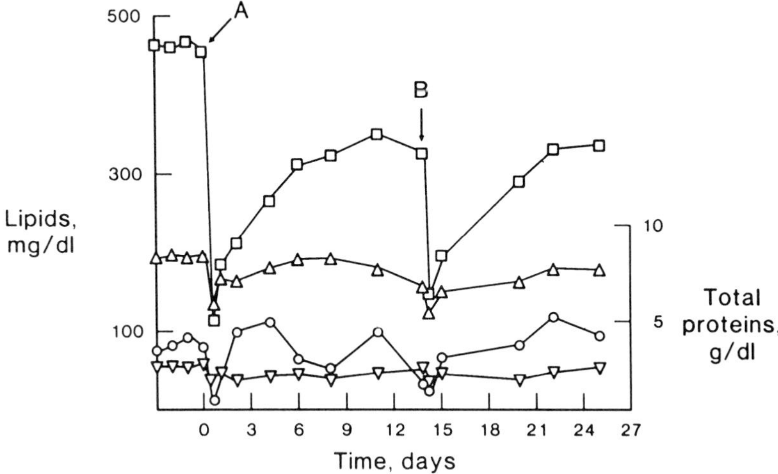

Figure 2: *Immediate and delayed effects of extracorporeal removal of LDL. A and B = consecutive treatments; □ = total cholesterol; △ = total protein; ○ = triglycerides; ▽ = HDL cholesterol.*

Table 3
Analysis of 139 Treatments in 6 Patients:
Changes in Various Blood Constituents

Constituent	*Change, %*
Hematologic:	
Hematocrit	−11.9
Hemaglobin	−11.4
MCV	None
RBC	−12.8
WBC	+53.0 (granulocytes incr.)
Platelets	−19.5
Electrolytes:	
Sodium	+ 1.1
Potassium	− 9.0
Calcium	− 7.4
Chloride	+ 2.7
PO$_4$	−12.7
Iron	−11.6
Enzymes:	
LDH	−15.1
GOT	−16.0
GPT	−17.0
γ-Glutamyl transferase	−16.4
Alkaline phosphatase	− 7.5
Other:	
Urea	−11.1
Uric acid	−13.8
Serum proteins:	
Total	−22.8
C3	−30.1
C4	−30.0
IgG	−22.5
IgA	−25.7
IgM	−30.9
Electrophoresis pattern	Unchanged

xanthomata in the interdigital webs of the hands was observed. Also, tuberous xanthomata over the olecranon and Achilles tendon shrank and softened considerably. This may be explained by an average loss of 130 g of LDL cholesterol over a 9-month period, which is about 64% more than has

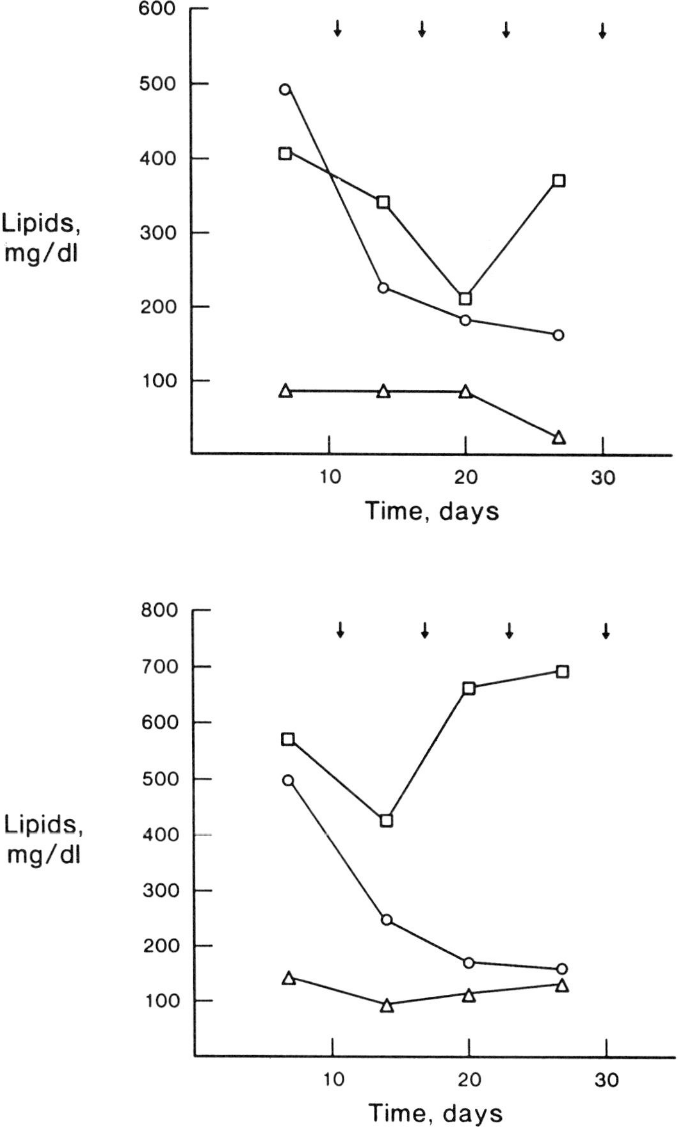

Figure 3: *Serum LDL cholesterol concentration and fecal neutral steroid and bile acid excretions during a 4-week cholesterol balance study in two heterozygous siblings. Daily concentrations were averaged over 1 week.* Upper: *in 14-year-old* girl; Lower: *in 12-year-old boy. Vertical arrows = treatments;* □ = *neutral steroids, mg/day;* ○ = *LDL cholesterol, mg/dl;* △ = *bile acids, mg/day.*

been reported for an alternative treatment.[31] The key question remains: Does the depletion of serum apo B-containing lipoproteins to a level well below normal achieve a regression of atherosclerosis? The subjective relief of angina pectoris, as well as electrocardiographic evidence (ST depression), suggests stabilization or regression of coronary atherosclerosis. Coronary angiography will be performed in the near future.

It must be stressed that the data presented represent the effect of a single therapy. There is little doubt that additional therapeutic intervention—e.g., drugs that interfere with the de novo synthesis of cholesterol—will further lower the LDL level.

SIDE EFFECTS

No side effects, such as fever, chills, syncope, or arrhythmias, were observed in any patient. Routine tests of blood and plasma revealed no chemical changes compared to the pretreatment period. Damage to erythrocytes or platelets, infections, or complement activation have not been observed. Patients were repeatedly tested by the passive anaphylaxis test[32] for human antisheep immunoglobulin antibody, but no sensitization was observed.

There is, however, the possibility of a minor decrease in blood pressure due to reduced blood volume during the phase when serum is running into the anti-LDL column while the saline of the dead volume is being discarded. This must be taken into consideration when treating children or persons with a reduced plasma volume.

LDL APHERESIS AND ALTERNATIVE TREATMENTS IN HOMOZYGOUS HYPERCHOLESTEROLEMIA

Because there is no functioning gene for the LDL receptor in the homozygous form of familial hypercholesterolemia, stimulation of receptor synthesis is ineffective. Until the early 1970s, no treatment was available. In 1973, portocaval shunt surgery was used to decrease plasma LDL levels in homozygotes.[33] Hepatic cholesterol synthesis and LDL synthesis were reduced 62% and 48%, respectively,[11] resulting in a reduction of plasma LDL cholesterol of about 40%. However, portocaval shunt surgery is not effective in all patients.[34] Severe side effects such as encephalopathy, hepatic injury,[35] and thrombosis caused by enhanced aggregation of thrombocytes have been reported.[36]

In 1975, plasma exchange was introduced as an effective therapy of homozygous hypercholesterolemia.[37] Replacement of 2 to 4.5 liters of plasma by fresh frozen plasma or plasma protein fractions with intervals of 3 weeks between treatments leads to a decrease in plasma cholesterol of about 50–65%.[30,31,38–41] The repeated depletion of the patient's plasma does not stimulate LDL synthesis.[29–31] On the contrary, evidence has been presented that body cholesterol pools are depleted. The effectiveness of the LDL removal, with relative absence of severe side effects, has made plasma exchange the preferred treatment in most medical centers.

The disadvantages of the procedure are obvious: it removes not only atherogenous lipoproteins from the patient's plasma but also other valuable plasma components—e.g., immunoglobulins and clotting factors. Plasma HDL concentration is also diminished,[37,39] which is unfavorable in view of the protective effect of this lipoprotein against the development of cardiovascular disease.[42]

The replacement of plasma by saline or dextran solutions (up to 1,000 ml) cannot be recommended because the resulting hypoproteinemia causes a secondary hypercholesterolemia[43] that would diminish the effect of the treatment. Fresh frozen plasma, used as a replacement fluid, contains cholesterol itself and may transmit serum hepatitis. Plasma protein fractions contain vasoactive substances—e.g., bradykinin—that may cause side reactions.[44] Plasma exchange therapy includes the necessity of continued treatment for an indefinite period, so the costs of the procedure are crucial factors. In the United States, the current cost is about \$800 per exchange,[31] which limits the application of this therapy to carefully selected patients.

Heparin-Sepharose affinity adsorption also has been tried in attempts to lower serum LDL levels. The results have been rather disappointing. A reduction of LDL cholesterol by only 20% could be achieved.[45,46] In addition to the relative ineffectiveness, the adsorbents are nonspecific. Numerous plasma proteins are known to have high binding affinity to immobilized heparin—e.g., antithrombin III and coagulation factors VII, IX, XI, XII, XIIa, and thrombin[47-52] components of the complement system[53] lipoprotein lipase,[54] and hepatic triglyceride lipase.[55] As in plasma exchange, HDL is also removed by heparin columns.[56]

In contrast to the approaches just mentioned, LDL apheresis constitutes a highly selective treatment for all forms of hypercholesterolemia. A 3-hour treatment removes up to 80% of plasma LDL. Besides LDL, only the apo B-containing VLDL is bound to a certain extent to the column. This is desirable because VLDL is the metabolic precursor of LDL. Other plasma components, especially HDL, are not retained. Immunoadsorption columns can be used for years without loss of binding activity. Because this therapy does not require human serum components, it constitutes an inexpensive alternative to plasma exchange.

Similar to plasma exchange, immunoadsorption has to be performed as a lifelong treatment with intervals of 7–14 days between sessions. In this respect, the selectiveness of the procedure seems especially valuable.

LDL apheresis represents the least expensive, most selective, and most effective treatment for familial hypercholesterolemia available at present. It has proved to be a safe procedure for decreasing LDL levels in homozygous patients, even to subnormal values. Whether or not this ultimately leads to a standstill, or even to a regression, of atherosclerosis is under investigation.

In principle, the combination of plasma separation and immunoadsorption in an extracorporeal circulation should be applicable to the selective removal of any plasma component of endogenous or exogenous origin, provided that the component has antigenic properties.

REFERENCES

1. Schettler G: Die Atiologie der Arteriosklerose. *Internist* (Berlin) 1978; 19:611.
2. Gotto AM Jr: Is atherosclerosis reversible? *J Am Diet Assoc* 1979; 74:551.
3. Castelli WP, Doyle JT, Gordon T, et al: HDL cholesterol and other lipids in coronary heart disease: The Cooperative Lipoprotein Phenotyping Study. *Circulation* 1977; 55:767.
4. Goldstein JL, Brown MS: The low density lipoprotein pathway and its relation to atherosclerosis. *Ann Rev Biochem* 1977; 46:897.
5. Kahn JA, Glueck CJ: Familial hypobetalipoproteinemia: Absence of atherosclerosis in a postmortem study. *JAMA* 1978; 240:47.
6. Glueck CJ, Gartside P, Fallat RW, et al: Longevity syndromes: Familial hypobeta and familial hyperalpha lipoproteinemia. *J Lab Clin Med* 1976; 88:941.
7. Fredrickson DS, Goldstein JL, Brown MS: The familial hyperlipoproteinemias. **In** *The Metabolic Basis of Inherited Disease.* Fourth edition. Edited by JB Stanbury, JB Wyngaarden, DS Fredrickson. New York, McGraw-Hill Book Company, 1978, p 604.
8. Goldstein JL, Brown MS, Stone NJ: Genetics of the LDL receptor: Evidence that the mutations affecting binding and internalization are allelic. *Cell* 1977; 12:629.
9. Brown MS, Goldstein JL: Lowering plasma cholesterol by raising LDL receptors. *N Engl J Med* 1981; 305:515.
10. Brown MS, Goldstein JL: Familial hypercholesterolemia: A genetic defect in the low density lipoprotein receptor. *N Engl J Med* 1976; 294:1386.
11. Bilheimer DW, Goldstein JL, Grundy SM, et al: Reduction in cholesterol and low density lipoprotein synthesis after portocaval shunt surgery in a patient with homozygous familial hypercholesterolemia. *J Clin Invest* 1975; 56:1420.
12. Langer T, Strober W, Levy RI: The metabolism of low density lipoprotein in familial type II hyperlipoproteinemia. *J Clin Invest* 1972; 51:1528.
13. Brown MS, Goldstein JL: The hyperlipoproteinemias and other disorders of lipid metabolism. **In** *Harrison's Principles of Internal Medicine.* Ninth edition. Edited by KJ Isselbacher, RD Adams, E Braunwald, et al. New York, McGraw-Hill Book Company, 1980, p 507.

14. Mordasini R, Riesen W: Pharmakotherapie der hyperlipidämien. *Ther Umsch* 1980; 37:990 (abstract).
15. Kovanen PT, Bilheimer DW, Goldstein JL, et al: Regulatory role for hepatic low density lipoprotein receptors in vivo in the dog. *Proc Natl Acad Sci USA* 1981; 78:1194.
16. Dietschy JM, Wilson JD: Regulation of cholesterol metabolism. *N Engl J Med* 1970; 282:1241.
17. Endo A, Kuroda M, Tanzawa K: Competitive inhibition of 3-hydroxy-3-methylglutaryl coenzyme A reductase by ML-236A and ML-236B fungal metabolites having hypocholesterolemic activity. *FEBS Lett* 1976; 72:323.
18. Alberts AW, Chen J, Kuron G, et al: Mevinolin: A highly potent competitive inhibitor of hydroxymethylglutaryl coenzyme A reductase and a cholesterol-lowering agent. *Proc Natl Acad Sci USA* 1980; 77:3957.
19. Yamamoto A, Sudo H, Endo A: Therapeutic effects of ML-236B in primary hypercholesterolemia. *Atherosclerosis* 1980; 35:259.
20. Thompson GR, Gotto AM Jr: Ileal bypass in the treatment of hyperlipoproteinaemia. *Lancet* 1973; 2:35.
21. Havel RJ, Eder HA, Bragdon JH: The distribution and chemical composition of ultracentrifugally separated lipoproteins in human serum. *J Clin Invest* 1955; 34:1345.
22. Noble RP: Electrophoretic separation of plasma lipoproteins in agarose gel. *J Lipid Res* 1968; 9:693.
23. Mancini G, Carbonara AO, Heremans JF: Immunochemical quantitation of antigens by single radial immunodiffusion. *Int J Immunochem* 1965; 2:235.
24. Grabar P, Williams CA: Méthode permettant l'étude conjuguée des propriétés électrophorétiques et immunochimiques d'un mélange de protéines: Application au sérum sanguin. *Biochim Biophys Acta* 1953; 10:193.
25. Stoffel W, Demant T: Selective removal of apolipoprotein B-containing serum lipoproteins from blood plasma. *Proc Natl Acad Sci USA* 1981; 78:611.
26. Mosbach K: Immobilized enzymes. *Methods Enzymol* 1976; 44:17.
27. Stoffel W, Borberg H, Greve V: Application of specific extracorporeal removal of low density lipoprotein in familial hypercholesterolaemia. *Lancet* 1981; 2:1055.
28. Stoffel W, Bode C, Borberg H, et al: Selective removal of plasma low density lipoproteins by combined extracorporeal plasma separation-immunoadsorption. **In** *Proceedings of the Sixth International Symposium on Atherosclerosis*, Berlin, 1982, p 502.
29. Soutar AK, Myant NB, Thompson GR: Metabolism of apolipoprotein B-containing lipoproteins in familial hypercholesterolaemia: Effects of plasma exchange. *Atherosclerosis* 1979; 32:315.
30. Berger GMB, Miller JL, Bonnici F, et al: Continuous flow plasma exchange in the treatment of homozygous familial hypercholesterolemia. *Am J Med* 1978; 65:243.

31. King MEE, Breslow JL, Lees RS: Plasma-exchange therapy of homozygous familial hypercholesterolemia. *N Engl J Med* 1980; 302:1457.
32. Ovary Z: Passive cutaneous anaphylaxis. *Meth Med* Res 1964; 10:158.
33. Starzl TE, Putnam CW, Koep LJ: Portocaval shunt and hyperlipidemia. *Arch Surg* 1978; 113:71.
34. Stein EA, Glueck CJ: Homozygous hypercholesterolemia treatment by portocaval shunt. **In** *Proceedings of the Fifth International Symposium on Atherosclerosis,* Houston, Texas, 1979. Edited by AM Gotto Jr, LC Smtih, B Allen. New York, Springer-Verlag, 1980.
35. Voorhees AB Jr, Chaitman E, Schneider S, et al: Portal-systemic encephalopathy in the noncirrhotic patient: Effect of portal-systemic shunting. *Arch Surg* 1973; 107:659.
36. Faergeman O, Gormsen J, Meinertz H: Anti-platelet drugs and portacaval anastomosis for homozygous hypercholesterolaemia. *Lancet* 1976; 2:1416 (letter to the editor).
37. Thompson GR, Lowenthal R, Myant NB: Plasma exchange in the management of homozygous familial hypercholesterolaemia. *Lancet* 1975; 1:1208.
38. Thompson GR: Plasma exchange for hypercholesterolaemia. *Lancet* 1981; 1:1246.
39. Thompson GR, Myant NB, Kilpatrick D, et al: Assessment of long-term plasma exchange for familial hypercholesterolaemia. *Br Heart J* 1980; 43:680.
40. Apstein CS, Zilversmit DB, Lees RS, et al: Effect of intensive plasmapheresis on the plasma cholesterol concentration with familial hypercholesterolemia. *Atherosclerosis* 1978; 31:105.
41. Witztum JL, Williams JC, Ostlund R, et al: Successful plasmapheresis in a 4-year-old child with homozygous familial hypercholesterolemia. *J Pediatr* 1980; 97:615.
42. Miller GJ, Miller NE: Plasma high density lipoprotein concentration and development of ischaemic heart disease. *Lancet* 1975; 1:16.
43. Lundsgaard-Hansen P: Intensive plasmapheresis as a risk factor for arteriosclerotic cardiovascular disease? *Vox Sang* 1977; 33:1.
44. Buskard NA: Plasma exchange and plasmapheresis. *Can Med Assoc J* 1978; 119:681.
45. Lupien P-J, Moorjani S, Lou M, et al: Removal of cholesterol from blood by affinity binding to heparin-agarose: Evaluation on treatment in homozygous familial hypercholesterolemia. *Pediatr Res* 1980; 14:113.
46. Lupien P-J, Moorjani S, Awad J: A new approach to the management of familial hypercholesterolaemia: Removal of plasma cholesterol based on the principle of affinity chromatography. *Lancet* 1976; 1:1261.
47. Miller-Andersson M, Borg H, Andersson L-O: Purification of anti-thrombin III by affinity chromatography. *Thromb Res* 1974; 5:439.
48. Kisiel W, Davie EW: Isolation and characterization of bovine factor VII. *Biochemistry* 1975; 14:4928.
49. Andersson L-O, Borg H, Miller-Andersson M: Purification and characterization of human factor IX. *Thromb Res* 1975; 7:451.

50. Fujikawa K, Thompson AR, Legaz ME, et al: Isolation and characterization of bovine factor IX (Christmas factor). *Biochemistry* 1973; 12:4938.

51. Koide T, Kato H, Davie EW: Isolation and characterization of bovine factor XI (plasma thromboplastin antecedent). *Biochemistry* 1977; 16:2279.

52. Fujikawa K, Kurachi K, Davie EW: Characterization of bovine factor XII$_a$ (activated Hageman factor). *Biochemistry* 1977; 16:4182.

53. Von Zeipel G, Hanson H-S, von Stedingk L-V: Purification from euglobbulin of the first component (Cl) of complement and its subcomponents by heparin-Sepharose chromatography. *Acta Pathol Microbiol Immunol Scand* [C] 1977; 85:123.

54. Iverius P-H, Östlund-Lundqvist A-M: Lipoprotein lipase from bovine milk: Isolation procedure, chemical characterization, and molecular weight analysis. *J Biol Chem* 1976; 251:7791.

55. Kuusi T, Kinnunen PKJ, Ehnholm C, et al: A simple purification procedure for rat hepatic lipase. *FEBS Lett* 1979; 98:314.

56. Burgstaler EA, Pineda AA, Ellefson RD: Removal of plasma lipoproteins from circulating blood with a heparin agarose column. *Mayo Clin Proc* 1980; 55:180.

CHAPTER 2

Selective Removal of Bile Acids

Alvaro A. Pineda, M.D.
Edwin A. Burgstaler, M.T.
E. Rolland Dickson, M.D.
Howard F. Taswell, M.D.

INTRODUCTION

Bile acids accumulate in tissues of patients with cholestasis,[1] and the unremitting pruritus that often accompanies cholestasis has been attributed to accumulation of bile acids in the skin.[2-7] Such pruritus is usually an extremely uncomfortable condition that can result in lack of sleep, skin ulcerations and excoriations, and mental depression. In approximately 20% of cholestatic patients, the secondary pruritus is refractory to conventional modes of treatment, which include corrective surgery and medications such as phenobarbital, cholestyramine, or activated charcoal capsules. When these have failed, various methods for removing pruritogenic substances from the circulating blood have been employed with varying degrees of success.

This chapter is a comparative review of three blood treatment methods for relief of intractable pruritus of cholestasis: (1) passage of plasma through charcoal affinity columns for removal of specific substances (this method will be represented by three examples), and nonselective removal therapy by (2) plasma exchange and by (3) hemodialysis. The comparisons have to do with the lowering of plasma concentrations of bile acids, the subjective response of patients treated, and the costs of treatment.

Because recent evidence refutes the idea of a direct causative role of bile acids in the genesis of pruritus,[8–11] alternative explanations are considered. One possibility is that the pruritogenic substance or substances travel in the company of bile acids. Alternatively, a placebo effect of the blood treatment methods is a possible explanation for the relief observed.

PLASMA PERFUSION OF CHARCOAL AFFINITY COLUMNS

Systems

Three charcoal affinity column systems for plasma perfusion treatment of intractable pruritus have been reported: the Mayo Clinic system,[12–15] the Cleveland Clinic system,[16] and the Nijmegen University system.[17]

Figure 1 depicts the plasma perfusion system used at the Mayo Clinic. Heparinized .citrated whole blood is pumped from the patient into a blood cell separator (either IBM or Haemonetics). Plasma is separated from the formed elements by centrifugal force and pumped at 40–60 ml/min by a peristaltic pump through the charcoal column. The 600 ml column contains powdered charcoal (coating 200 μm glass beads, to which it is held by polyvinylpyrrolidone).[12] From the column, the plasma passes through a 1.5 μm filter to trap any free charcoal and then is recombined with the formed elements in a reservoir from which the treated whole blood is returned to the patient. When the IBM Cell Processor is used, the peristaltic pump and the reservoir are eliminated.

Figure 2 depicts the plasma perfusion system used at the Cleveland Clinic. Whole blood is heparinized and pumped through a cellulose acetate hollow fiber plasma filter (Ashahi

Plasma Perfusion of Affinity
Columns or Filters with Immobilized Sorbent

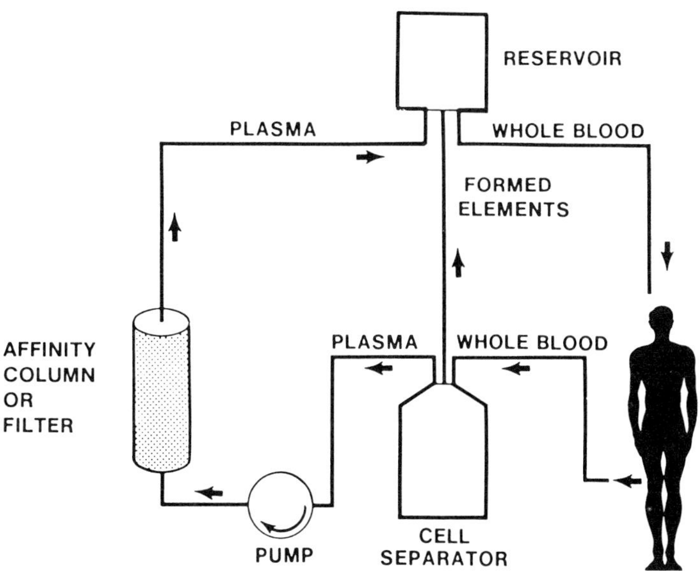

Figure 1: *Plasma perfusion system used at Mayo Clinic. (Used with permission of publisher and author.)*[13]

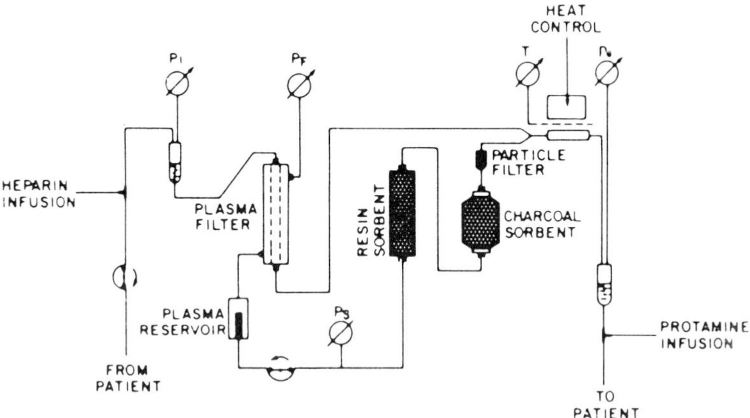

Figure 2: *Plasma perfusion system used at Cleveland Clinic. (Used with permission of publisher and author.)*[3]

Medical Co., Tokyo). The filter separates the plasma from the formed elements, and the plasma passes into a reservoir. A second pump is used to pump the plasma through an anionic exchange resin cartridge, then through a cartridge containing coated activated charcoal (Hemosorba, Ashahi Medical Company, Tokyo). The treated plasma then passes through a filter to remove particles and is recombined with the formed elements, and the treated whole blood, after rewarming, is returned to the patient.

Figure 3 depicts the plasma perfusion system used at Nijmegen University. Anticoagulated blood is pumped from the patient into a blood cell separator (Celltrifuge), where plasma is separated from the formed elements by centrifugal force. The plasma is pumped at 30 ml/min through commercially built activated charcoal columns (Becton Dickinson, Gambro, and Haemocol adsorbers). To increase bile acid charcoal transfer, a recycling system is attached to attain a plasma flow rate of 200 ml/min through the adsorber. By means of another pump, the treated plasma is drawn from the reservoir, mixed with the formed elements, and returned to the patient.

Bile Acid Removal

Table 1 shows the amount of experience with each of the three charcoal affinity columns described and presents the

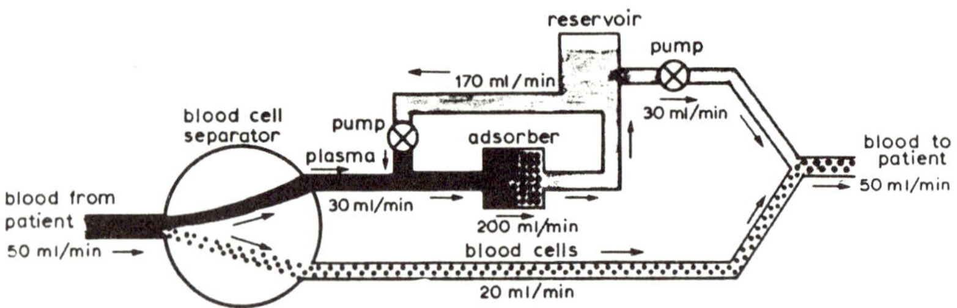

Figure 3: *Plasma perfusion system used at Nijmegen University. (Used with permission of publisher and author.)*[6]

Table 1
Plasma Perfusion of Charcoal Affinity Column
Data from the Three Institutions

	Mayo Clinic	Cleveland Clinic	Nijmegen University
Patients, no.	25	6	1
Perfusions, no.	168	27	5
Volume perfused, liters: mean	4.0		
range	1.0–6.0	1–7	2.7–5.4
Duration, hours	2.0–3.5	4–8	1.5–3.0
Bile acids, mean μmol/liters			
Before perfusion	198	NA	143
After perfusion	138	NA	129
(% change)	(30)	(40)*	(10)
Bile acids removed, μmol: mean	545	150*	202
range			150–315
Column efficiency (% removal)	(99)	(98)*	(66)

*Cholyglycine only; NA = not available.

data obtained. The mean volume of plasma perfused was similar in the three institutions—approximately 4 liters. The mean time required for a perfusion, however, was more than twice as long at the Cleveland Clinic as at either of the others.

The Mayo Clinic system removes more of the total bile acids per perfusion than the other systems. Note that the bile acid data listed for the Cleveland Clinic in Table 1 relate to only the cholyglycine fraction. On that limited basis, the Cleveland Clinic system removes a greater percentage than the other systems, but the absolute quantity it removes is the least. The columns used by the Mayo Clinic and Cleveland Clinic were very efficient, accomplishing 99% and 98% removal, respectively. However, the columns used by the Nijmegen University had an efficiency of 66%.

The crucial accomplishment, of course, is removal of the pruritogenic substance from the skin tissue. We have calculated that perfusions using the Mayo Clinic system mobilized an average of 368 (SD 147) μmol of bile acids from extravascular compartments. Comparable calculations for the other systems have not been reported.

Clinical Experience (Mayo Clinic)

Patients and Management

Table 2 summarizes relevant clinical and laboratory data from the 25 Mayo Clinic cases. There were 19 women and six men. Their ages ranged from 16 to 80 years, with a median of 46 years. The diagnosis was primary biliary cirrhosis in 15 cases, primary sclerosing cholangitis in four, benign recurrent cholestasis in two, and one case each of psoriasis with probable cryptogenic cirrhosis, undetermined liver disease with portal hypertension (Alagile syndrome, and carcinomatosis with obstructive jaundice and liver metastasis). The patients had significant cholestasis: total bilirubin averaged 10.7 mg/dl (SD 8.0) and alkaline phosphatase was 1,802 U/liters (SD 1,150).

All of the patients had unremitting pruritus that interfered seriously with sleep and daily activities and led to varying degrees of skin excoriation secondary to scratching. Plasma perfusion of the activated charcoal-coated glass bead column was undertaken to alleviate these effects. The number of treatments per patient ranged from one to 27, with a median of seven. (The technical data are in Table 1.)

Clinical Response

Of the 25 patients, 20 (80%) reported a beneficial therapeutic effect: complete absence in four and nearly complete in 10, marked amelioration in four, and slight in two. No relief was obtained by five, including two who received two courses of three plasma perfusions each. One of the unresponsive patients had psoriasis, which was believed to contribute to the pruritus.

The relief following a course of two or three plasma perfusions was reported to last from 24 hours to 12 months, averaging 9 weeks. One patient reported 2 years of complete relief following a single plasma perfusion. In addition to the subjective relief reported by patients, there was objective evidence: reduced excoriations, increased sleep hours unaided by barbiturates, and ability to wear customary clothing.

Because of the simplicity of the procedure, the therapeutic effect perceived by the patients and the absence of adverse effects, the procedure was accepted very well by all patients. Of the 25 treated, 13 returned for additional perfusions and three had 18 or more perfusions over a period of 2 years.

Side Effects and Minor Reactions

With charcoal affinity column perfusion, as with any medical treatment, side effects are an important consideration. With the exception of a single moderate hypotensive episode in one patient and a severe unexplained allergic reaction in another patient, there were no significant adverse effects either during or after plasma perfusion using the Mayo Clinic system. Occasionally, however, the procedure induced a drop in blood pressure or chilling of the patient (effects seen with plasma exchange procedures also). In addition, some patients have experienced episodes of nausea and lightheadedness and mild citrate toxicity.

The symptoms of hypovolemia usually can be controlled by administration of fluids and are less frequent with use of continuous flow cell separators than with the semicontinuous type. Chilling of the patient can be prevented and also treated by use of blood warmers on the return lines of the perfusion equipment and blankets on the patient. Citrate toxicity can be diminished to some degree by reducing the citrate dose used.

In one Mayo Clinic case, a moderate allergic reaction followed a plasma perfusion. This reaction—swelling of the eyelids—occurred after the patient had left the Apheresis Laboratory. Recently the same patient has been treated a second and third time under premedication with steroids and diphenhydramine (Benadryl), and at the end of these proce-

Table 2
Plasma Perfusion of Charcoal Affinity Column: Clinical and Laboratory Data from Mayo Clinic Cases

Case No.	Age, Sex	Wt, Kg	Diagnosis	Total bili, mg/dl	Alk Phos, U/liter	Perfusions No.	Courses	Mean ml	BA, µmol/liter Before perf.	After perf.	BA, µmol Retained by column	Mobilized from EVC	Clinical Results
1	44 F	49	Primary biliary cirrhosis	17.0	5034	2	1	4323	338	191	847	466	Almost complete relief for 2 wk; gradual return of pruritus over 4 wk, much relief. Pruritus of genital lichen planus did not change.
2	64 M	80	Primary sclerosing cholangitis	2.6	1382	3	1	4094	129	54	192	8	Almost complete relief for 1 wk, followed by slight amelioration for 2 wk.
3*	52 F	66	Primary biliary cirrhosis	11.0	2176	7	2	3451	315	209	847	513	No improvement.
4*	38 F	59	Primary biliary cirrhosis	7.0	2750	3	1	3921	328	248	865	625	Almost complete relief for 8 wk.
5	26 F	59	Sclerosing cholangitis	10.0	1746	27	9	3688	346	244	663	312	Almost complete relief for 5 wk, followed by marked amelioration for 9 wk.
6	53 F	54	Primary biliary cirrhosis	1.4	1130	3	1	3582	234	201	587	498	Marked amelioration for 6 wk, slight amelioration for 3 wk more.

7	42 M	65	Primary sclerosing cholangitis	13.2	3200	6	2	4044	225	161	560	363	No improvement.
8	80 M	57	Psoriasis. Probable cryptogenic cirrhosis	0.8	224	3	1	4012	58	50	178	139	No improvement. Patient had psoriasis.
9*	40 M	71	Primary sclerosing cholangitis	16.1	2675	5	1	3806	203	167	533	403	Slight amelioration for 4 wk.
10	35 M	79	Indeterminate liver disease with portal hyper-tension	0.6	442	3	1	3752	48	60	525	508	Complete relief for 2 wk, almost complete relief for 2 wk more.
11*	48 F	41	Primary biliary cirrhosis	14.7	871	3	1	3543	284	183	544	355	Slight amelioration for 3 wk.
12	48 F	45	Primary biliary cirrhosis	16.0	1066	2	1	4750	60	95	566	566	Marked amelioration for 14 wk.
13	50 M	64	Benign recurrent intrahepatic idiopathic cholestasis	8.3	734	13	3	4610	157	126	566	450	Almost complete relief for 5 wk.
14	39 F	52	Primary biliary cirrhosis	8.0	1368	5	2	5033	251	150	699	430	Complete relief for 24 hr; gradual return of pruritus.

(Continued)

Table 2 (Continued)

Case No.	Age, Sex	Wt, Kg	Diagnosis	Total bili, mg/dl	Alk Phos, U/liter	Perfusions No.	Perfusions Courses	BA, µmol/liter Mean ml	BA, µmol/liter Before perf.	BA, µmol/liter After perf.	BA, µmol Retained by column	BA, µmol Mobilized from EVC	Clinical Results
15*	63 F	63	Primary biliary cirrhosis	3.3	978	7	3	5194	131	94	424	291	Almost complete relief for 6 wk after first course. Marked amelioration for 2 wk after second course.
16	66 F	41	Primary biliary cirrhosis	0.7	1219	1	1	2980	48	43	417	408	Complete relief for 2 wk; then 9 mo of marked amelioration.
17	21 F	64	Benign recurrent cholestasis	32.0	609	6	2	4068	266	159	539	170	Almost complete relief for 9 wk.
18	54 F	68	Malignant carcinoid with obstructive jaundice and liver metastasis	20.3	830	2	1	5102	103	55	408	235	No improvement.
19*	52 F	47	Primary biliary cirrhosis	8.0	2516	2	1	1552	195	154	226	119	Marked amelioration for 9 wk.
20	39 F	49	Primary biliary cirrhosis	12.0	3163	8	3	3581	261	192	520	337	Almost complete relief for 10 mo.
21*	32 F	55	Primary biliary cirrhosis	17.5	1727	5	2	4136	192	130	549	377	Marked amelioration for 9 wk.

22	65 F	64	Primary biliary cirrhosis	1.9	1553	3	1	5000	119	50	600	399	No improvement. (Responded to hypnosis.)
23	16 F	27	Alagile syndrome	6.0	2912	4	2	2825	205	126	490	366	Complete relief for 1 yr.
24*	39 F	58	Primary biliary cirrhosis	18.0	3400	18	6	4597	256	175	759	505	Almost complete relief for 9 wk; then marked amelioration for 6 wk following first 3 courses of treatment. Last 2 courses gave slight amelioration.
25	50 F	53	Primary biliary cirrhosis	21.5	1340	27	13	4654	189	125	522	346	First courses produced almost complete relief for up to 5 wk; subsequent courses induced marked amelioration for 9 wk and less as clinical condition deteriorated.

*Deceased. Bili = bilirubin; Alk phos = alkaline phosphatase; BA = bile acids; perf = perfusion; EVC = extravascular compartments.

dures she has experienced mild to moderate hoarseness and difficulty in swallowing.

The Nijmegen University report did not mention any side effects of the perfusions. The Cleveland Clinic system, like the Mayo system, occasionally induced hypotension or chilling.

Pruritogenic Substances

The prompt response in 20 of 25 patients treated suggests that a substance causing pruritus equilibrates readily between the skin tissues and the intravascular compartment and is removed by plasma perfusion of the charcoal column. Possibly this pruritogenic substance (or substances) travels in the company of bile acids, which are efficiently removed by the column. But alternatively, a placebo effect is a possibility. In the absence of a therapeutic alternative, charcoal column plasma perfusion is a safe and reasonably effective modality. There is great need, however, for controlled trials, especially because of the subjectiveness of the effect sought, namely, control of pruritus. The difficulty lies in obtaining acceptance of the long, mildly unpleasant sham procedures by patients who are already experiencing great discomfort.

Cost

One unavoidable consideration is that of cost. Charcoal affinity column perfusion has an advantage over plasma exchange in not needing replacement fluid, but the equipment to separate plasma from the formed elements is still needed. Until recently, this required centrifugal force separators, which involve an initial equipment outlay of $20,000 to $30,000. With the introduction of new filtration systems, this initial outlay should be significantly less. Anyway, most institutions planning to use the charcoal perfusion system already have purchased a cell separator for other procedures.

It is difficult to give the cost of the filters and other consumables. The columns used by the Mayo Clinic are prepared by the staff of the Apheresis Laboratory. For each perfusion, the material cost per perfusion is approximately $50, and preparation of the column takes 3½ hours of technicians' time. If commercial columns become available in the future, it will be easier to estimate the true cost of a procedure. Large-volume production of columns should reduce costs.

PLASMA EXCHANGE

Although much has been published about plasma exchange as therapy for several diseases, there are very few papers on its use in treatment of intractable pruritus.[17,18] One study of this treatment for pruritus has been reported from Nijmegen University[17] and one from Rayne Institute.[19] Once vascular access has been established, anticoagulated whole blood is pumped into a separator (either centrifugal or filtrational) where plasma is separated from the formed elements.

The patient's plasma is collected in transfer packs and discarded. The volume of plasma removed is replaced with combinations of isotonic saline, albumin solution, fresh frozen plasma, or plasma protein fraction. The replacement fluids are mixed with the formed elements at the same rate that the plasma is removed, and the reconstituted whole blood is returned to the patient. The procedure is continued until 0.5 to 2.0 times the patient's plasma volume has been exchanged.

Table 3 gives data obtained from plasma exchange as treatment for intractable pruritus: 26 exchanges in two cases at Nijmegen University and 31 exchanges in three cases at Rayne Institute. The amount exchanged at Nijmegen University was 2−3 liters, which would be approximately one plasma volume of an adult of average size. This is twice the amount exchanged at Rayne Institute. The procedure times were not given, but at the Mayo Clinic a plasma exchange procedure for 1.5 plasma volumes requires 2.5 to 3.5 hours. Bile acid measurements were not made at the Rayne Institute.

In the Nijmegen University experience, plasma exchange removed a mean of 480 μmol of bile acids for treatment, lowering the circulating concentration by a mean of 20%. This is comparable to the results obtained by the Mayo Clinic charcoal affinity column. Clinically, the two Nijmegen University patients had intractable pruritus with severe cholestatis, which was secondary to primary biliary cirrhosis in one case and in the other was secondary to extrahepatic biliary obstruction caused by a pancreatic carcinoma. Three exchanges were performed before amelioration of pruritus occurred; then weekly exchanges sustained the therapeutic effect. The treatment was considered palliative. Slight adverse effects of treatment were reported.

The group from Rayne Institute reported similar results. They treated three patients with primary biliary cirrhosis and intractable pruritus who were not responsive to conventional therapy, which included cholestyramine, norethindrone, and phenobarbital. In addition, the patients had hypercholesterolemic xanthomas, and one had angina pectoris. In one case, the plasma exchanges reduced cholesterol levels and size of xanthomas but did not control pruritus. In the second case angina ceased, some cutaneous lipid deposits disappeared, cholesterol concentrations diminished, and pruritus gradually lessened, becoming much less severe after 8 months of periodic exchanges. In the third case, itching stopped for 3 weeks after each plasma exchange but returned rapidly to intolerable levels after 3 weeks.

Plasma exchange induces numerous side effects,[18] which usually are related to hypovolemia, citrate toxicity, or reaction to replacement fluids. The Rayne Institute reported hypovolemia symptoms in two of the three patients treated. Nijmegen University reported shaking episodes in patients receiving plasma protein fraction. The measures previously recommended for management of hypovolemia and citrate toxicity with charcoal affinity column treatment could also be used with plasma exchange. Allergic reactions to replacement fluids may be minimized by using more crystalloid fluid or albumin solution as replacement fluid, since they contain no protein or a single protein as potential allergen.

Unfortunately, there are no reports of controlled trials of plasma exchange in cholestasis, not to mention its effect on pruritus. Costs for plasma exchange vary considerably, depending on the type of equipment used to remove the plasma, the volume of exchange, and the type of replacement fluid used. Because of the cost of replacement solutions, plasma exchange would be more expensive than plasma perfusion of charcoal columns.

HEMODIALYSIS

Hemodialysis has been used to treat pruritus of cholestasis in one case reported from the Wilhelmina Gasthuis University Hospital.[20] The cholestasis, secondary to chronic nonsuppurative cholangitis, was severe. The pruritus was intense and had proved resistant to cholestyramine and phenobarbital. The dialysis used an artificial kidney with a 1-m^2 polyacrylonitode membrane (Rhone-Povlenc) and a single-path system. Blood flow was approximately 300 ml/min and dialysate flow was 500 ml/min. The duration of the procedure averaged 6 hours. Thirty treatments were given to the one patient (Table 3). Dialysis reduced the pretreatment concentration of bile acids by a mean of 19% per procedure and the mean total of bile acids removed was 666 μmol.

There was an acceptable correlation between the bile acid level at the end of dialysis and the intensity of pruritus as measured by a score devised by the authors. This study involved the use of a control. Of the 30 procedures performed, six were shams in which the bathing fluid was conveyed

Table 3
Plasma Exchange and Hemodialysis

	Plasma Exchange		Dialysis
	Nijmegen University	*Rayne Institute*	*Wilhelmina Gasthuis Univ. Hosp.*
Patients, no.	2	3	1
Treatments, no.	26	31	30
Volume, exchanged or dialyzed, liters	2−3	0.67*	108
Duration, hr.	NA	NA	6
Bile acids, mean μmol/liters:			
Before perfusion	236	NA	200
After perfusion	189	NA	162
(% change)	(20)	NA	(19)
Bile acids removed μmol: mean	480	NA	666
range	245−640		

*Proportion of patient's plasma volume (not liters).

around (rather than through) the artificial kidney; and interestingly enough, alleviation of pruritus also occurred with the sham dialyses. This strongly suggests a placebo effect of the treatment or a therapeutic alteration of blood caused by extracorporeal circulation. No side effects were reported in the patient. One would expect the same as those that occur with dialysis in kidney failure.

Comparisons: The volume of blood processed in the dialysis system is much greater than that passed by affinity columns or plasma exchange, but the efficiency of bile acid removal is less than that of plasma exchange and much less than that of plasma perfusions and charcoal affinity columns. These results are very close to those obtained with the Mayo Clinic charcoal affinity column and the Nijmegen University plasma exchange.

The cost savings of hemodialysis over charcoal affinity columns and plasma exchange depend on the number of dialysis procedures required to obtain relief from pruritus. Three dialysis procedures would cost less than three charcoal affinity columns or three plasma exchanges. If more than three dialysis procedures are required, the cost advantage diminishes.

COMPARISON OF METHODS

Of the reports reviewed, all but that from the Mayo Clinic dealt with a small number of patients. This makes drawing conclusions on each procedure very difficult. After consideration of the data presented, selection of a method probably will depend on the availability of equipment and personnel. Institutions currently treating diseases with plasma exchange and not interested in preparing or purchasing affinity columns may choose to treat pruritus with plasma exchange. On the other hand, institutions that currently perform hemodialysis, but not plasma exchange, may consider the use of hemodialysis to treat pruritus if further studies document a beneficial effect. However, the advantages of selective removal therapy with affinity columns should not be overlooked.

Charcoal columns are more specific in their removal capabilities than either plasma exchange or hemodialysis, as indicated by postperfusion normal chemistry levels. The plasma perfusions of charcoal affinity columns need no replacement fluids such as those required with plasma exchange. This decreases the cost and the chance of reaction to the replacement fluid. For the slower flow rate of plasma perfusing the affinity column, antecubital venipuncture usually is adequate, rather than the cannulation required for hemodialysis. This is important, because most patients will receive intermittent treatment as the need occurs.

The data indicate that the Mayo Clinic charcoal affinity column accomplishes bile acid removal and pruritus relief with fewer hours of treatment than the other methods. Also, in regard to the amount of total bile acids removed per procedure, the Mayo Clinic system is clearly superior to other charcoal affinity column systems. However, the average bile acid removal for the Mayo Clinic charcoal column is very similar to the average removal demonstrated in the Nijmegen University plasma exchange and the hemodialysis performed at Wilhelmina Gasthuis University Hospital. The papers reviewed spoke of the great need for controlled studies. As suggested by the sham hemodialysis, a placebo effect may influence the response of the patient being treated for intractable pruritus.

Since actual cost was not given for the procedures reviewed, cost comparison is not possible. As an indication, we can say that at the Mayo Clinic three charcoal affinity column procedures cost about 65% as much as a plasma exchange (including replacement fluids) and 136% as much as a hemodialysis procedure.

PRURITOGENIC AGENT

One of the biggest problems in assessing the effect of treatment for intractable pruritus of cholestasis is identification of the pruritogenic agent. At present, the best indicator as to effectiveness of treatment is the subjective response of the patients to their itching. The identification and measurement of the pruritogenic substance (or substances) would give greater validity to claims of therapeutic success. As previously mentioned, bile acids have been cited as a possible cause of pruritus.[15] Indeed, the reduction of serum concentrations of bile acids by the Cleveland Clinic charcoal affinity column and by the hemodialysis study correlates with relief of pruritus. On the other hand, studies have been cited that found no correlation of serum bile acid levels and degree of pruritus.[8-11]

Ghent and Bloomer advanced additional data to support the premise that pruritus in hepatobiliary disease is not related directly to bile acid retention.[9] They proposed that the pruritogenic agent is produced in response to cholestasis, possibly through activation of the alternate pathway of bile acid synthesis. The production and excretion of this agent are affected by therapy with cholestyramine, phenobarbital, estrogens, androgens, and possibly thyroxine. One possible pruritogenic agent could be porphyrins, since porphyrins have been associated with pruritus. Preliminary findings at the Mayo Clinic indicate a possible correlation between reduction of porphyrin concentration and relief of pruritus. Intractable pruritus is also found in renal disease;[21] there is a possibility that the same pruritogenic agent could be the cause of pruritus in both cholestasis and renal disease.

CONCLUSION

Further use of these methods for treatment of intractable pruritus of cholestasis should be restricted to patients who are unable to obtain relief with surgical or medical treatments. If the pruritogenic agent can be identified, the use of these methods and new procedures can be compared in larger studies. The use of these methods should be governed by the degree of relief they give the patient who has no other form of treatment available.

REFERENCES

1. Stiehl A: Gallensäuren und Gallensäurensulfate in der Haut von Patienten mit Cholestase und Juckreiz. *Z Gastorenterol* 1974; 12:121.
2. Ahrens EH Jr, Payne MA, Kunkel HG, et al: Primary biliary cirrhosis. *Medicine* (Baltimore) 1950; 29:299.
3. Carey JB Jr, Williams G: Relief of the pruritus of jaundice with a bile acid sequestering resin. *JAMA* 1961; 176:432.
4. Kirby J, Heaton KW, Burton JL: Pruritic effect of bile salts. *Br Med J* 1974; 4:693.
5. Schoenfield LJ: The relationship of bile acids to pruritus in hepatobiliary disease. **In** *Bile Salt Metabolism.* Edited by L. Schiff, JB Carey, Jr, J Dietschy. Springfield, Illinois, Charles C Thomas, 1969, p 257.
6. Schoenfield LJ, Sjövall J, Perman E: Bile acids on the skin of patients with pruritic hepatobiliary disease. *Nature* 1967; 213:93.
7. Varco RL: Intermittent external biliary drainage for relief of pruritus in certain chronic disorders of the liver. *Surgery* 1947; 21:43.
8. Freedman MR, Holzbach RT, Ferguson DR: Pruritus in cholestasis: No direct causative role for bile acid retention. *Am J Med* 1981; 70:1011.

9. Ghent CN, Bloomer JR: Itch in liver disease: Facts and speculations. *Yale J Biol Med* 1979; 52:77.
10. Ghent CN, Bloomer JR, Klatskin G: Elevations in skin tissue levels of bile acids in human cholestasis: Relation to serum levels and to pruritus. *Gastroenterology* 1977; 73:1125.
11. Summerfield JA, Bartholomew C, Billing BH, et al: Bile acid profiles in skin interstitial fluid and serum: Relationship to pruritus. *Gastroenterology* 1980; 79:1126 (abstract).
12. Lauterburg BH, Dickson ER, Pineda AA, et al: Removal of bile acids and bilirubin by plasma perfusion of USP charcoal-coated glass beads. *J Lab Clin Med* 1979; 94:585.
13. Lauterburg BH, Pineda AA, Burgstaler EA, et al: Treatment of pruritus of cholestasis by plasma perfusion through USP charcoal-coated glass beads. *Lancet* 1980; 2:53.
14. Lauterburg BH, Pineda AA, Dickson ER, et al: Plasma perfusion for the treatment of intractable pruritus of cholestasis. *Mayo Clin Proc* 1978; 53:403.
15. Pineda AA, Burgstaler EA: Unpublished data.
16. Carey WD, Smith J, Asanuma Y, et al: Pruritus of cholestasis treated with plasma perfusion. *Am J Gastroenterol* 1981; 76:330.
17. Geerdink P, Snel P, van Berge Henegouwen GP, et al: Treatment of intractable pruritus in patients with cholestatic jaundice by plasma exchange and plasma perfusion. *Neth J Med* 1978; 21:239.
18. Pineda AA: Therapeutic applications of plasma exchange and cytapheresis. *Clin Lab Annu* 1983; 2:145.
19. Keeling PWN, Bull J, Kingston P, et al: Plasma exchange in primary biliary cirrhosis. *Postgrad Med J* 1981; 57:433.
20. Hoek FJ, Grijm R, Sanders GTB, et al: Removal of bile acids from the blood by hemodialysis with a polyacrylonitril membrane: Treatment of pruritus of cholestatic disease. *Digestion* 1982; 23:135.
21. Silverberg DS, Iaina A, Reisin E, et al: Cholestyramine in uraemic pruritus. *Br Med J* 1977; 1:752.

CHAPTER 3

Selective Removal of A and B Isoagglutinins

William I. Bensinger, M.D.

INTRODUCTION

Interest in removing anti-A and anti-B antibodies from plasma initially grew out of a practical need to prevent hemolysis during marrow infusion in ABO-incompatible marrow transplant recipients. However, it became clear that the removal of anti-A and anti-B antibodies could serve as a model system for the manipulation of plasma that would have applicability to other problems. We have been interested in the removal from plasma of other antibodies and substances such as antiplatelet antibodies, anti-HLA antibodies, myeloma proteins, factors associated with hepatic coma, immune complexes, etc. It appeared that the model of red cell antibody removal was the best available to remove selectively an antibody without removing other substances. We reasoned that if

This investigation was supported by grants CA 18579 and CA 18029 awarded by the National Cancer Institute, DHHS.

we were successful in a single system, then all we would need to proceed with removal of other substances would be to have the appropriate antigenic or reactive substance properly immobilized on a column. This chapter describes our work on removal of anti-A and anti-B antibodies and clearly demonstrates that once a specific immunoadsorbent is available, the application will proceed very rapidly.

HISTORICAL PERSPECTIVE

It was Landsteiner in 1901 who first described the presence of naturally occurring antibodies (isoagglutinins, now referred to as alloagglutinins) in human serum.[1] These antibodies were directed against antigens found on the surface of red cells. Two types of antigens were discovered by differential agglutination that segregated into four blood types based on whether cells from a particular person carried one antigen, the other antigen, both antigens, or none. These antigens, termed ABO, are simple carbohydrate chains. A and B antigens differ only by the presence, as the terminal sugar, of N-acetylgalactosamine and galactose, respectively. These antigens are determined by transferases which catalyse the transfer of sugar to the oligosaccharide chains already on the cells. The transferases are the products of the genes of the ABO system.[2] Group O cells generally contain only H substance, but this is also found on A, B, and AB cells as it is the basic oligosaccharide antigen which is converted to A or B antigen by the transferases. Only the rare Bombay blood type has no H antigen.

Most humans of blood group A produce anti-B antibodies. Group B humans produce only anti-A. Group O humans produce both anti-A and anti-B antibodies. It is usually only persons of blood group AB who are unable to produce anti-A or anti-B antibodies.

Some persons may have naturally occurring antibodies to A and B red cells. The exact reason for this is unknown;

however, one potential stimulus is the presence of intestinal bacteria, which have antigens cross-reactive with A and B antigens.[3] Pregnant women may be sensitized by placental blood leakage from an ABO-incompatible fetus.

Red cell transfusions to human subjects must be ABO-"compatible"; that is, A, B, or AB cells cannot be given to patients of another blood group having the anti-A or anti-B antibody. However, A cells or B cells may be given to people of blood group AB. O cells may be given to anyone. AB plasma or other blood group plasma containing low A and B antibody titer may be transfused to anyone. High titered blood group O, A, or B plasma may be given only within the respective blood groups.

A and B antigens are sometimes found on platelets and white cells but appear to be adsorbed antigens rather than manufactured by the cell.[4] ABO antigens have been found on other tissues including endothelial cells, spermatozoa, pancreas, and gastric mucosa. However, it is still controversial as to whether these antigens are produced by the cell or simply adsorbed from plasma.[5,6] This may be a relevant issue regarding organ transplantation since the continued production of anti-ABO antibody by specific host tissues could jeopardize survival of an incompatible graft.

ORGAN GRAFTING

Initial organ grafting was performed between identical twins where compatibility of all antigens was assured.[7] The first allogeneic human organ transplants were performed more than 25 years ago using kidneys grafted into patients with chronic renal failure.[8] Success rates were low, however, and it was another 10 years before kidney grafting was considered a routine treatment for chronic renal failure. Red cell antigens were not thought to contribute to the allograft rejection process. However, most kidney transplants performed were between ABO-compatible individuals, that is, group A,

B or AB kidneys were not given to recipients possessing the corresponding red cell antibody.

Hume et al.[8] reported a single ABO-incompatible transplant in 1955 that failed. It was 8 years later that Starzl et al. reported their results with two ABO-incompatible kidney grafts.[9] His initial results were good with two surviving grafts. However, more extensive experience in ABO group-incompatible renal transplants revealed several instances of hyperacute rejection.[10] It was therefore assumed that major ABO mismatching was a contraindication for renal transplantation;[11] this was especially true when live donors were involved. Later work indicated this was probably not true; however, this belief became established in renal transplant centers worldwide. This may have been partially due to the change from exclusively live donors to the use of cadaver kidneys when a wider range of kidneys became available.

The first ABO-incompatible marrow transplant was performed in 1971.[12] Anti-A antibody was successfully reduced in the patient by the intravenous infusion of Witebsky blood group A substance and A red cells. Although not known at that time, the presence of anti-red cell antibody would not have affected graft function since ABH antigens are not found on marrow precursor cells.[13] Gale et al. later reported successful engraftment of marrow from blood group A or B donors into patients with anti-A or anti-B antibodies.[14] The number of patients studied was small, but there did not appear to be any undue incidence of graft rejection or graft-versus-host disease.

Gale et al. used two primary methods for dealing with anti-A or anti-B antibody present in the patients prior to transplant: large volume plasma exchange and in vivo antibody absorption. Plasma exchange was achieved using a continuous flow cell centrifuge (Celltrifuge, American Instrument Company) and replacement with 15−20 liters of plasma isovolumetrically. In vivo absorption of antibody was achieved by the transfusion of incompatible donor-type red cells or the infusion of group A substance prepared from hog stomach.

We studied ABO-incompatible marrow transplantation in Seattle beginning in 1972. Seventeen patients were initially treated for major ABO incompatibility.[15] Anti-red cell antibody titers ranged from 1:8 to 1:2048. Antibody was removed by plasma exchange or reconstituted whole blood made with recipient-type red cells and donor-type plasma. All 17 patients engrafted. There was one possible graft rejection in the patient with the highest antibody titer pretransplant. There was no greater incidence of graft-versus-host disease or other complications in this group of patients. It was thus confirmed that marrow transplantation could be performed across major ABO incompatibility.

The Seattle group has had a more extensive experience with ABO-incompatible marrow grafts. We have now performed transplants in 56 patients with major ABO incompatibility. Fifty-one patients had sufficiently high anti-red cell antibody titers to require large volume plasma exchange of 5−30 liters over 1−3 days. This was followed by an exchange of 2−4 units of incompatible whole blood. Five patients with low antibody titers simply had whole blood exchange. Donor and recipient blood group types are shown in Table 1.

These patients had an outcome that appeared to be comparable to a large group of similar but ABO-compatible transplant recipients. The incidence of graft-versus-host disease, interstitial pneumonia, graft rejection, infection, and leukemic relapse was similar. Of interest was that in a group of 12

Table 1
ABO-Incompatible Transplants

Donor	Recipient	No.
A	O	30
B	O	10
B	A	4
A	B	3
AB	B	3
AB	A	6
	Total	56

ABO-incompatible, HLA-matched marrow transplants performed for patients with aplastic anemia, there was only one rejection. It may be possible that plasma exchange facilitates engraftment by the addition or removal of soluble factors other than anti-ABO antibodies.

One of the clinical problems following ABO-incompatible marrow transplantation is the return of circulating isoagglutinins. This problem was extensively analyzed in a larger series of patients transplanted in Seattle through 1980.[16] In 81 ABO-incompatible marrow transplant recipients, significant antibody returned in 29 patients. There was evidence that this was newly synthesized antibody and that it was both IgM and IgG. Antibody return correlated with pretransplant antibody levels and was unaffected by the intensity of transplant conditioning; that is, cyclophosphamide plus total body irradiation was no more effective in suppressing posttransplant antibody production than cyclophosphamide alone.

However, the return of antibody did not affect time to successful marrow engraftment, graft rejection, graft-versus-host disease, or posttransplant blood product requirements. Furthermore, antibody levels disappeared by 100 days posttransplant. These above findings indicate that the primary usefulness of selective removal of anti-red cell antibody prior to ABO-incompatible marrow transplantation was to avoid hemolysis at the time of marrow infusions.

While plasma exchange clearly facilitated the successful engraftment of ABO-incompatible marrows, there are a number of drawbacks to this approach. Firstly, whole plasma exchange is a relatively gross way to remove only a small amount of a single antibody from patients. Large volume plasma exchange is very costly with estimates ranging from $1,000 to $2,000 per treatment just for the plasma. Albumin solutions appear to cost at least as much and will deplete patients of other important plasma constituents such as clotting factors. Patients undergoing plasma exchange frequently have adverse reactions, including allergic symptoms, volume overload, or citrate toxicity. Finally, the use of large volume plasma exchange requires plasma from a great many patients and will thus increase the risk of transmission of hepatitis or other virus-related infections.

Because of these problems, a method of specific removal of anti-red cell antibody would be preferable. Our first approach was to develop a column of red cell ghosts, but this was impractical.[17] One logical approach was to develop an antigen affinity column of either purified or synthesized blood group antigen chemically linked to a solid support. Whole blood or separated plasma could then be passed over such a column. Ideally, the perfect column would be biocompatible with either whole blood or plasma, would require minimal anticoagulation of the patient, and would selectively remove the desired anti-A or anti-B isoagglutinin without removing other plasma components. The A and B immunoadsorbent columns tested at the Fred Hutchinson Cancer Research Center are close to achieving these goals.

Measurement of Antibodies to ABO Blood Group Antigens

Human antibody to A and B red cell antigens is measured by in vitro agglutination. Antibodies of IgM class is measured by doubling tube dilutions of serum to be tested added in equal amounts to a 2% red cell suspension. Samples are incubated at room temperature for 15 minutes and red cell agglutination is scored usually with the aid of a microscope. The highest tube dilution capable of producing agglutination of type-specific red cells is taken as the titer of antibody.

For measurement of IgG class antibodies, serum is first incubated with Cleland's sulfhydryl reagent for 15 minutes at 37°C.[18] This process disrupts the disulfide bonds of IgM antibodies, rendering them incapable of agglutination. Doubling tube dilutions of the treated serum are then added to the type-specific red cells and incubated for 45 minutes at 37°C.

After centrifuging the cells and decanting the supernatant, the cells are washed three times with saline to remove non-specifically bound immunoglobins. An antibody to human immunoglobulin (indirect Coombs reagent) is then added to the cells, centrifuged, and again scored for agglutination.

The limitation of the in vitro measurement of anti-red cell antibody is the poor correlation of the level of antibody with the potential for a hemolytic transfusion reaction when incompatible blood is given. This calls into question the usefulness of measuring antibody removal since both plasma exchange and immunoadsorption leave some residual antibody behind. This issue will be addressed in more detail when the topic of efficiency of immunoadsorption is discussed.

Synthetic A and B Blood Group Antigens

The achievement of making a clinically useful immunoadsorbent capable of removing anti-A and anti-B antibodies was in large part due to the pioneering efforts of R.U. Lemieux. He developed a technique for the chemical synthesis of oligosaccharide antigenic determinants of a number of important blood group substances.[19] It then became feasible to synthesize on a commercial scale certain blood group antigens that are monospecific haptens of their carbohydrate antigenic determinants. They were used for several years as in vitro immunoadsorbents by bonding these synthetic haptens chemically to an insoluble solid matrix. The solid immunoadsorbent was then used in affinity chromatography to purify and concentrate specific antibodies from immune serum.

Because these in vitro immunoadsorbents met many of the specifications required for the ideal clinically useful immunoadsorbent, we undertook a project for the development of an immunoadsorbent that could be applied to the treatment of patients where it was desirable to remove specifically anti-red cell antibodies.

Testing of A and B Immunoadsorbents

Initial testing of the immunoadsorbent columns was used to establish effectiveness and specificity of antibody removed and as a gauge to the quantity of immunoadsorbent needed to remove antibody from patients. In initial work, we established an optimum bead size for the insoluble matrix that would permit the high plasma flows required of a clinical immuno-adsorbent.

For the in vitro testing, 2 liters of pooled blood bank human group O plasma containing high levels of anti-A and anti-B antibody were circulated through an immunoadsorbent column of synthetic human blood group A. The column contained 50 g of synthetic A immunoadsorbent. The circuit was recirculated into the flask containing the plasma to try to simulate conditions that would exist in patients. The plasma flow rate was 30 ml/minute. Samples for anti-A and anti-B antibody were obtained hourly for anti-red cell determination.

Results from a typical experiment are depicted in Figure 1. As shown, there is a rapid and specific decline in the level of anti-A antibody, both IgM and IgG, within 3 hours. In contrast, levels of anti-B antibody changed little although there was a small decline in anti-B IgM. This effect is well known and represents the removal of antibody with specificity for both A and B red cells that blood group O people are known to make.[20] These experiments demonstrated that it was feasible to use relatively compact columns (50−100 g) to remove antibody from patients in clinically desirable situations.[21]

Another in vitro experiment was performed in order to assure that the immunoadsorbent columns were removing anti-A or anti-B antibody instead of merely neutralizing the antibody in the plasma by the leaching of antigen off the column. Two and one-half milligrams of affinity chromatography-purified human anti-A IgG was radiolabeled with I^{125} and diluted into 2 liters of pooled human group O plasma with low anti-A titer. The plasma was distributed into two equal volumes of 1

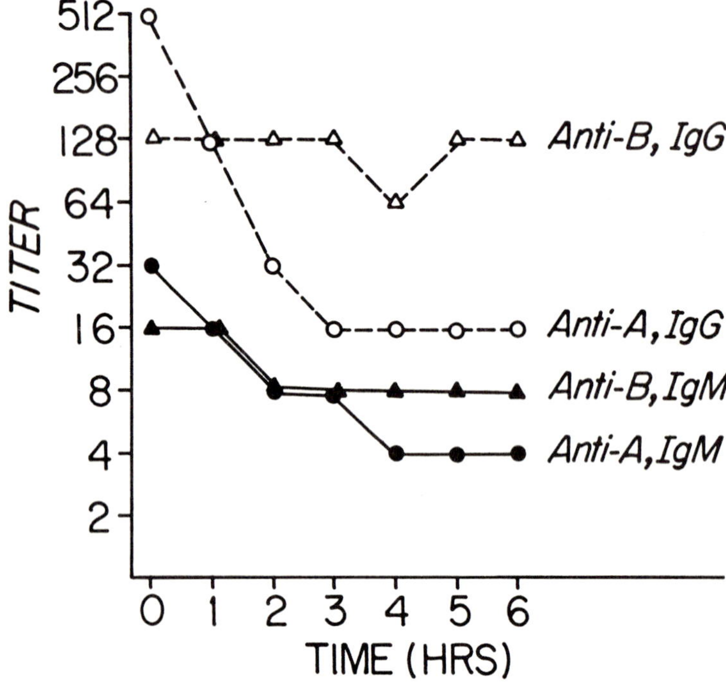

Figure 1: *In vitro testing of synthetic A immunoadsorbent. Two liters of pooled group O plasma were recirculated over a 50-g immunoadsorbent column at 30 ml/min for 6 hours. Samples were obtained hourly from the pooled plasma for anti-A and anti-B determinations.*

liter each and circulated for 6 hours over 2 g of A antigen-linked silica. The results are depicted in Figure 2. It can be seen that approximately one-half of the radioactivity was removed from the plasma circulated over the A antigen column but there was no change in activity of the plasma from the control, blank silica column. At the termination of the experiment, the whole columns were counted by gamma scintillation. Only background radioactivity was recorded by the silica while almost 1,000 times more counts accumulated on the A-linked antigen column, demonstrating uptake of labeled antibody by the immunoadsorbent column.[21]

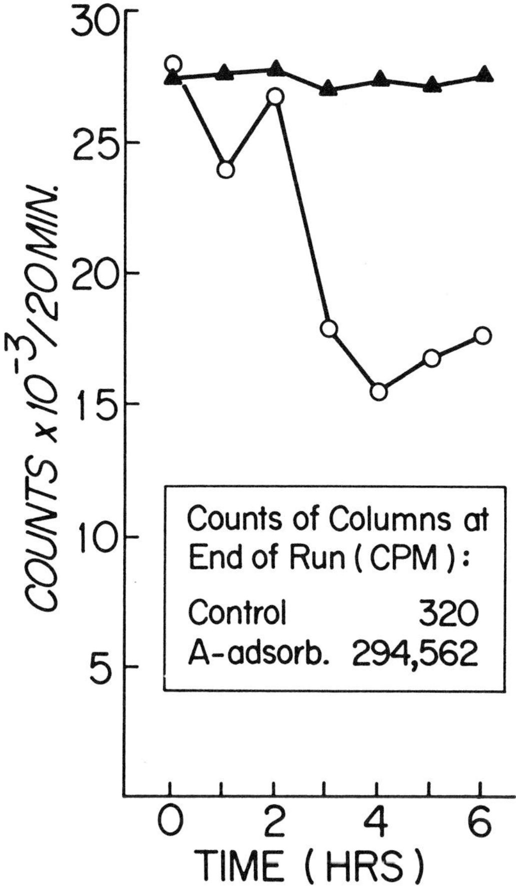

Figure 2: *I^{125}-labeled anti-A diluted in low titer group O plasma divided into two equal volumes and circulated over a synthetic A column or a control column containing blank silica. Samples were obtained hourly for radioactivity. The whole columns were counted at the end of the procedure. O—O—O = radioactivity of synthetic A circuit. ▲—▲—▲ = radioactivity of control circuit.*

Animal Studies

Before clinical trials in humans could begin, it was necessary to demonstrate efficacy without undesirable toxicity. We chose to work with dogs because they were large enough to allow the significant extracorporeal blood volume needed for a blood cell separator and column apparatus. Although the dog is a relatively difficult model to work with because of the fragile red cells, it was felt that success with an immunoadsorbent column in this model would probably assure success when studies were advanced to humans.

Most animal sera contain heteroagglutinins that react with red cells from other animals, including humans. Thus, in order to detect anti-A or anti-B antibody by red cell agglutination techniques, it is first necessary to perform tedious absorptions with group O cells. A more suitable technique utilized I^{125}-radiolabeled synthetic human blood group A to measure anti-A antibody. This was a modification of the technique of Minden and Farr which was used to measure antibodies to bovine serum albumin.[22] Dogs were then actively immunized to synthetic human blood group A and bovine serum albumin as a control antibody. After 8–14 weeks the dogs were used for in vivo immunoadsorption experiments. Teflon-Silastic shunts were placed between the carotid artery and jugular vein in the dogs' necks. The dogs were anticoagulated with heparin 100 units/kg and connected via the shunt to a continuous flow centrifuge (American Instruments Company, Silver Spring, Maryland). The centrifuge separated whole blood into plasma and formed elements by centrifugation at 1,500 g. Separated plasma was then pumped over a 50–g synthetic A immunoadsorbent column at 25–30 ml/minute and returned to the dog after recombination with formed elements of the blood. Dogs were treated for 2.5–3 hours and had two to three plasma volumes perfused through the columns.

Data from three dogs are presented in Figure 3. There was a specific decline in the levels of anti-A antibody with little or no change in the level of antibovine serum albumin antibody in all the dogs. The dogs received no immunosuppressive therapy, and levels of anti-A antibody typically rebounded during the posttreatment phase. Our assay technique did not permit us to distinguish between IgM and IgG

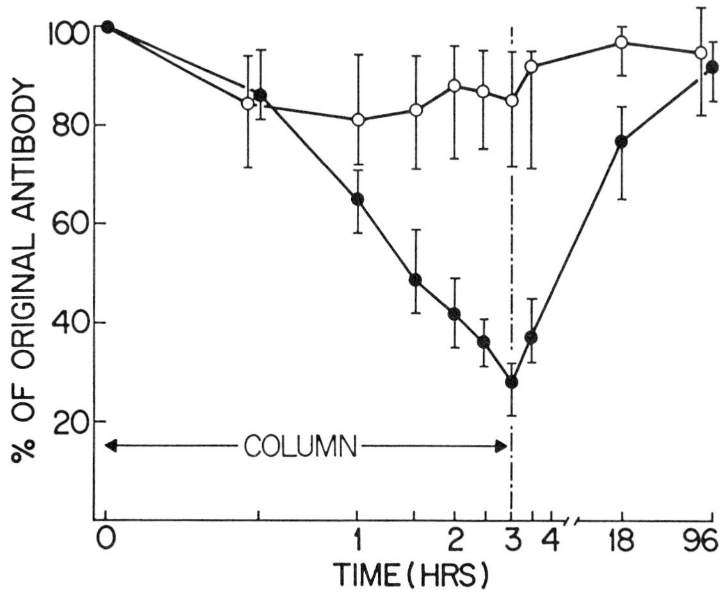

Figure 3: *In vivo testing of synthetic A immunoadsorbent. Three actively immunized dogs were connected to a blood cell separator and the plasma was pumped over a 50-g immunoadsorbent column for 3 hours. Samples from the dog were obtained every 30 minutes for anti-A and anti-B determinations. ●—●—● = anti-A. ○—○—○ = anti-BSA. Points represent the mean values for the three dogs with bars representing the range of values. (Used with permission from the publisher.)[21]*

antibody specific for A antigen. We were thus unable to determine whether the rebound was due to reequilibration of antibody from extravascular sites or to synthesis of new antibody. However, the rapid time course suggests reequilibration.

Further studies using a second, new column added to the cell separator in another dog showed that not only could additional anti-A antibody be removed but that the immediate rebound effect could be attenuated. It may thus be feasible to change columns in the middle of a procedure in order to remove more antibody.

The immunoadsorption procedure was remarkably well tolerated by the dogs with few adverse effects.[21] No changes in pulse rate or respiratory rate were noted. Platelet counts declined an average of 40% with each procedure. There were no major changes in hematocrit or white blood cell count. Most serum chemistries did not change before or after treatment. Free hemoglobin did increase an average of 80 mg/dl in the dogs but was undetectable by 12 hours posttransplant. We found this was due to red cell destruction at the seal interface of the centrifuge bowl and not due to any effect of the immunoadsorption columns. The dogs showed no adverse effects from the free hemoglobin, especially on renal function.

Protocol for Use of Immunoadsorption in Humans

Synthetic A and B antigens are very stable when covalently bound to the solid-bed silica support. They may be kept in a dry form for months. Several different sterilization methods can be used without loss of activity. Ten percent formaldehyde treatment for 1 hour or dry heat at 120°C for 24 hours may be used with less than 10% loss of activity. The preferred method is to expose the cartridges to ethylene oxide at 55°C for 4 hours at a pressure of 10 pounds per square inch followed by aeration for 12 hours. After dry sterilization the columns are stable for several months.

The day before use the columns are wetted with 1 liter of isotonic saline. Then 500 ml of 1% human serum albumin is added and the cartridge is incubated at 4°C for 12 hours or overnight. Just prior to use the column is again washed with 1 liter of isotonic saline containing 5,000 units of heparin and 0.2% sodium citrate.

The immunoadsorption system is depicted schematically in Figure 4. Plasma separation may be effected with a membrane device, continuous or intermittent flow blood cell separator. The separated plasma is then circulated through the immunoadsorption cartridge and rejoined to the formed blood elements.

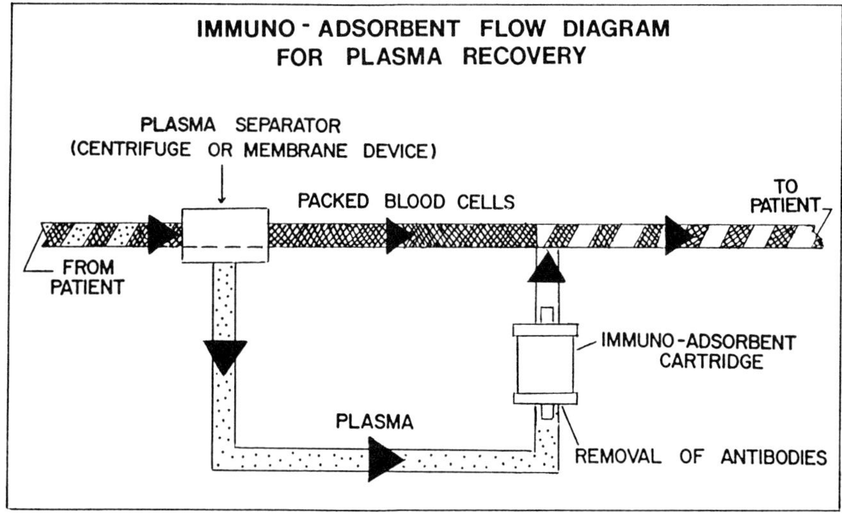

Figure 4: *Schematic diagram depicting the arrangement of the immunoadsorption system.*

A typical immunoadsorption system is used as follows. Patients are anticoagulated with 2,000−4,000 units of sodium heparin intravenously. They are connected to a blood cell separator for plasma separation. Anticoagulation is continuous using a heparin concentration of 40 units/ml and 3% sodium citrate. The anticoagulant flow rate is adjusted as necessary from 1−3 ml/minute. Separated plasma is pumped over the immunoadsorbent columns at 25−40 ml/minute. Patients have two to three calculated plasma volumes perfused over the column. Plasma volumes are estimated in liters on the basis of 5% of body weight in kilograms.

The columns as used above are somewhat sensitive to clotting when there is a high level of platelet contamination in the separated plasma. This appears to be due to heparin's aggregating effect on platelets. In such situations it may be necessary to use extra citrate in the anticoagulant when treating patients with high initial platelet counts (i.e., patients with chronic myelogenous leukemia). Preliminary work using citrate as the sole anticoagulant is very promising since there appear to be no problems with clotting or clogging of the columns. At the end of the procedure, the plasma in the columns may be returned to the patient by displacement with isotonic saline containing 15 units/ml of heparin.

We have performed studies on the feasibility of reusing the A and B immunoadsorption columns.[23] It is possible to regenerate the columns after use by the elution of bound proteins using high molality salt solutions or chaotropic ions.

Human Studies

The results of the anti-A and anti-B immunoadsorption studies in vitro and in vivo using dogs were quite successful. These initial studies were encouraging enough that clinical trials of specific immunoadsorption were begun.

The logical treatment group for this study included patients undergoing ABO-incompatible marrow transplantation at the Fred Hutchinson Cancer Research Center. Eleven patients initially had immunoadsorption performed using 80-g immunoadsorbent columns containing either synthetic A or B antigen.[23]

Patients were treated in a fashion very similar to the dog immunoadsorption studies except for the following modifications. The initial studies used Teflon-Silastic shunts placed in the forearm for all procedures. Two models of blood cell separators were used for the studies (Aminco or IBM 2997), with plasma circulated through the columns at 30 ml/minute for 4 hours. Effective removal of antibody occurred in all 11 patients. The columns reduced anti-A or anti-B antibody by an average of three doubling tube dilutions, representing 80−85% of intravascular antibody if reequilibration from extravascular sites is ignored.

The procedure appeared to be specific since anti-B antibody titers did not change in five group O patients undergoing immunoadsorption for removal of anti-A antibody. While the procedure appeared to remove the bulk of anti-A and anti-B antibodies, there was almost never complete removal of specific antibody. This is probably due to the kinetics of any exchange or adsorption procedure which can never remove 100% of a substance but may also be related to the narrow specificity of the synthetic antigen.

As in the dogs, there was an average 47% decline in platelet counts after treatment. This is similar to the losses we observe in patients undergoing large volume plasma exchange and probably represents platelet damage from the centrifugation technique. The immunoadsorbent columns clearly do not remove platelets from the patients since the measurement of column inlet and column outlet platelet counts shows no change. The columns may, however, be responsible for the "activation" of platelets passing through with their subsequent consumption or destruction in the patient.

There were no significant changes in serum chemistries measured before and after treatment. In contrast to the dog studies, there was no hemolysis and no free hemoglobin detected in the plasma of patients after immunoadsorption.

The success of specific plasma immunoadsorption for anti-A and anti-B antibodies led to studies of a larger group of patients undergoing immunoadsorption.[24] We have now treated 46 patients with synthetic A or B antigen columns for removal of specific antibody and performed 67 immunoadsorption procedures. The results of these treatments are presented in Table 2. For simple analysis of data, the doubling tube dilution results were converted to percent antibody re-

Table 2
Immunoadsorption for Removal of Anti-A or Anti-B Antibody

	Treatment 1			Treatment 2		
	No. of Treatments	IgM Percent	IgG Antibody Removed	No. of Treatments	IgM Percent	IgG Antibody Removed
Anti-A	30	73	75	13	72	69
Anti-B	17	80	79	7	81	80

moved. As seen in Table 2, synthetic A and B antigen columns successfully removed approximately 70–80% of intravascular antibody. This is equivalent to a two plasma-volume plasma exchange, assuming no extravascular mixing. These data include a few patients treated in whom immunoadsorption was relatively ineffective at reducing anti-A or anti-B antibody titers.

Although the immunoadsorption procedures are generally efficient, there are occasional situations in which the immunoadsorbent is ineffective in removing antibody. The first series of problems is seen in Figure 5 A and B. In Figure 5A, a patient with a relatively high titer of anti-B (IgM 1:256, IgG 1:1024) has good removal of the specific antibody. Monitoring the column effluent plasma, "post" (open circles) shows that even after 4 hours the column is capable of clearing additional antibody from the plasma. In contrast, the patient in Figure 5B has a relatively low initial anti-A titer (IgM 1:16, IgG 1:8), yet after 3 hours the column appears to be "saturated," that is, column effluent plasma antibody titers equal the column inlet titers. While this could be a true saturation phenomenon, this is unlikely since patient 1178 in Figure 5A has a much larger quantity of antibody than patient 1155 in Figure 5B, even when size differences are considered. More likely, this phenomenon relates to the specificity of anti-A or anti-B antibody produced by each patient. Patient 1178 has a larger proportion of "anti-B" antibody reactive with the synthetic group B determinant on the column. Patient 1155, in contrast, has relatively little antibody reactive with the synthetic A determinant on the immunoadsorbent, yet this patient does have antibody reactive with natural A determinants on red cells. The presence of polymorphic components of blood group ABH determinants is well known.[25] This highlights one of the drawbacks of using synthetic monospecific antigens for immunoadsorption of naturally occurring antibodies. While most patients make anti-A or anti-B antibody reactive with the monospecific antigens, a few do not, and this may lead to poor responses to immunoadsorption.

A few other problem patients came to light during treat-

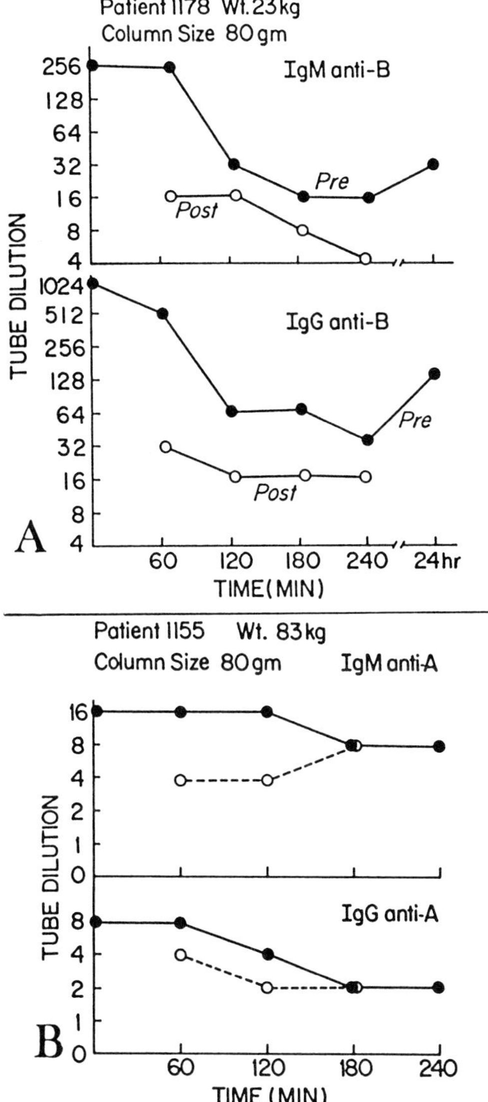

Figure 5 A,B: *Immunoadsorption of anti-B antibody from a patient prior to ABO-incompatible marrow transplant. "Pre" refers to plasma antibody levels drawn from the patient. "Post" refers to plasma antibody levels drawn from the outlet side of the immunoadsorbent column. (Used with permission of the publisher.)*[24]

ment of a larger series of patients. One patient with aplastic anemia with relatively high preadsorption titers of anti-A (IgM 1:1000, IgG 1:500) had only slight decreases in titers to IgM 1:500 and IgG 1:256 after the treatment. Further treatments with new columns failed to produce a decrement in antibody titers. When the serum from this patient was analyzed, it was found to bind well to the synthetic A immunoadsorbent at 4°C but very poorly at 37°C which is near the operating temperature of the columns. Another patient had an anti-A antibody of IgM specificity that bound well to the synthetic immunoadsorbent and an IgG anti-A that bound poorly. On examination, the IgG anti-A had a Forssman antigen specificity but still bound well to group A red cells.

These findings underscore the relative limitations of specific antigens used for immunoadsorption of naturally occurring antibodies. They do raise another issue when the immunoadsorption procedure fails to clear completely the antibody from plasma. The "safe" level of anti-A or anti-B antibody whereby one can transfuse incompatible red blood cells is unknown. Our studies indicate that levels of 1:16 or less, either IgG or IgM, are associated with no risk of a hemolytic transfusion reaction when incompatible red blood cells are given. It is unclear whether higher levels of antibody are unsafe. We have safely infused whole marrow into a single patient who had an anti-A IgG titer of 1:32 18 hours after immunoadsorption. Another patient with an anti-A titer of 1:16 IgM and 1:64 IgG was inadvertently given blood group A marrow without hemolysis.

It may be that the immunoadsorbent columns remove the more dangerous antibody (the higher affinity antibody) even when they are not completely effective at reducing anti-A or anti-B titers. However, this hypothesis remains to be proven.

Alternate Methods of Dealing with ABO-Incompatible Marrows

Other techniques described for dealing with incompatible marrow utilize red cell depletion to avoid hemolytic transfusion reactions. These methods use gravity sedimentation with hydroxyethyl starch or centrifugation in an intermittent or continuous flow blood cell separator to deplete the marrow of red cells.[25] These methods are effective in avoiding hemolysis during and after marrow infusion. Since only late red cell precursors express the ABO blood group antigens,[13] the presence of anti-A or anti-B antibody should not harm stem cells.

The major drawback to these techniques is that manipulation of marrow always results in loss of cells. While there is no reliable in vitro assay for the pluripotent stem cell, measurement of CFU_C (a committed granulocyte precursor cell) before and after processing marrows has shown an average loss of 47%.[25] The loss of large quantities of stem cells may or may not be important when transplanting patients with leukemia who are treated with chemotherapy and irradiation. However, in patients with aplastic anemia, cell dose has been shown to be an important determinant of engraftment.[26] Thus, any procedure that results in lowered stem cell doses may be undesirable in such cases.

Kidney Transplantation

As mentioned previously, most transplant surgeons consider major ABO incompatibility between donor and recipient to be an absolute contraindication to renal transplantation.[10, 11,27] Evidence for such a belief is scant at best. Indeed, Slapak et al. have described the successful long-term engraftment of a kidney from a group A donor to a group O recipient.[28] Although the patient experienced an episode of acute rejection during the early phase of the transplant, it was reversed by a modified plasmapheresis and immunoadsorption technique.

Plasma obtained from the patient during the initial exchange was batch-absorbed with group A red cells. The treated plasma was then used for the subsequent exchange. The patient received six daily exchanges of 3.5 liters using red cell-adsorbed autologous plasma. Reversal of the rejection correlated with a drop in the anti-A titer. This patient is now more than 3 years posttransplant with a functioning graft. Success has also been reported in grafting kidneys from A_2 donors into recipients of blood group 0.[29]

The above studies are consistent with the published but largely ignored studies of Wilbrandt et al.[30] who analyzed the ABO blood group incompatibility problem. They found that of 12 major ABO-incompatible transplants, only three survived long term. However, when anti-AB alloagglutinin titers were measured during the posttransplant phase, all three patients with surviving grafts had low antibody titers in contrast to three patients from the nine unsuccessful grafts who had antibody titers measured and had very high anti-AB titers. The antibody titers posttransplant from the other six unsuccessful transplants were not reported.

The presence of A and B blood group antigens on kidney endothelial cells has been described.[6] What is not known is whether these antigens are produced by the cells or whether they are merely adsorbed from plasma in a manner analogous to platelets. This issue is important since it may be possible to keep antibody levels reduced by immunoadsorption in the recipient of an ABO-incompatible graft for several weeks. This may allow sufficient time for shedding of adsorbed antigens on the cell surface. Alternately, endothelial cells that make A or B antigen could undergo antigenic modulation in the presence of low levels of antibody. This may allow the host to become "tolerant" of an ABO-incompatible kidney. Wilbrandt's data suggest this may be possible.[30] It is clear that this is an important area of investigation and should be actively pursued.

Other Blood Group Antigens in Kidney Transplantation

An analysis of factors responsible for graft rejection showed that Lewis-positive kidneys have a significantly lower success rate when given to Lewis-negative recipients.[31] This problem was magnified when Lewis-negative recipients having anti-Lewis antibody were given kidneys from Lewis-positive donors.[32]

The antigenic structure of the Lewis blood group is well known, and it is possible to construct antigen columns capable of removing Lewis antibody through immunoadsorption. Whether removing Lewis antibody from kidney graft recipients prior to transplant of a Lewis-positive kidney will affect graft survival is unknown but is worthy of study.

WHOLE BLOOD IMMUNOADSORPTION

The technique of plasma immunoadsorption requires a continuous flow blood cell separator or similar device for separation of plasma to be treated. These machines are expensive and require a specially trained operator for their use.

Immunoadsorbent systems compatible with whole blood increase the utility and reduce the expense of the procedure. Work using rabbits demonstrated the feasibility of removing specific antibody by hemoperfusion. As a further refinement of the technique of immunoadsorption of anti-A and anti-B antibody, we have developed a method of whole blood hemoperfusion of the specific A antigen columns that is quite effective in dogs.[33]

Synthetic human blood group A immunoadsorbent covalently bonded to crystalline silica was coated with collodion (cellulose nitrate) dissolved in an ether:alcohol mixture in a ratio of 3:40 after a modification of Chang.[34] Columns were sterilized with ethylene oxide as previously described.

Dogs actively immunized to bovine serum albumin and human blood group A were used for the hemoperfusion studies. Four immunized dogs had Teflon-Silastic shunts placed between the carotid artery and jugular vein for the studies.

We used systemic anticoagulation consisting of an intravenous heparin (100 units/kg) bolus. Continuous anticoagulant consisting of 3% sodium citrate and heparin 10 μ/ml was mixed with blood in a blood:anticoagulant ratio of 10:1. The procedures lasted 2½ hours. The four dogs underwent a total of six immunoadsorption procedures with two dogs having a second treatment sufficiently far apart in time to allow recovery of antibody levels.

Antibody levels to bovine serum albumin were measured as previously described. Anti-A antibody was measured by a solid-phase immunoadsorbent assay.[33]

Results of five procedures are shown in Figure 6, which shows that anti-A antibody levels declined rapidly during whole blood immunoadsorption. There was essentially no change in control antibovine serum albumin antibody levels after allowing for initial hemodilution. The sixth procedure was done using a previously used immunoadsorbent column that was regenerated and resterilized. Significant activity remained, although removal of anti-A was somewhat less effective.

Platelet counts measured before and after treatment showed declines that were quite variable. The average loss was 46% which is comparable to plasma immunoadsorption or plasma exchange where loss of platelets averaged 50%. We were concerned that platelets lost might be aggregating on the immunoadsorbent surface. To check for this, the used immunoadsorbent material was examined by electron microscopy. No platelet or fibrin material was detected.

Total hemolytic complement (CH_{50}) measured pre- and posttransplant showed about a 50% decline. Serum was examined for visible hemolysis before and after treatment. There was no evidence of visible hemoglobin in any samples. This is a significant finding since dog red cells are known to be fragile and lyse easily. Overall, the procedures were well tolerated. About one-half of the time the dogs experienced citrate toxicity with tetany; however, they responded promptly to intravenous calcium. We believe this toxicity was due to the small size of the dogs (10–19 kg) and the relatively high blood flow rates (30–50 ml/minute) that required a considerable amount

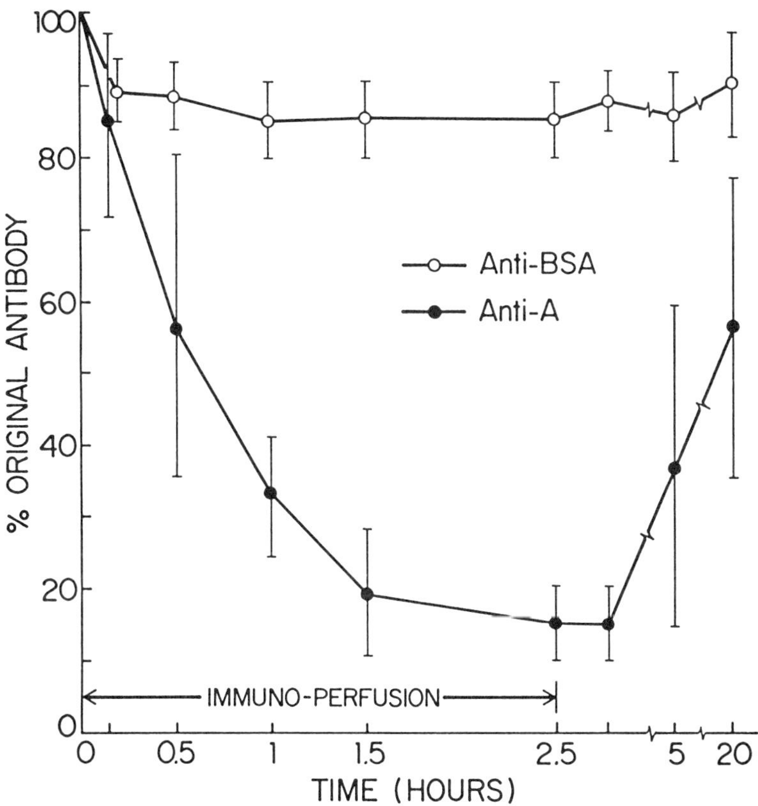

Figure 6: *Whole blood immunoadsorption of anti-A antibody. Dogs actively immunized to A antigen and BSA antigen had whole blood perfusion over 50-g synthetic A columns for 2½ hours. Data points represent the mean of the three dogs with bars showing the range. (Used with permission of the publisher.)*[33]

of citrate to be infused. There were no long-term sequelae from the treatment in any dogs.

The results in dogs were quite promising. We were impressed that specific immunoadsorption could be accomplished with whole blood, that no severe clotting problems developed, and that red cells did not appear to be affected by hemoperfusion.

We have thus extended these studies to our patient population undergoing ABO-incompatible marrow transplantation. The conditions of treatment have been similar to those used in the dog studies. We have studied four patients so far with results equal to immunoadsorption with plasma. No side effects have been observed. There has been no free hemoglobin in the plasma post treatment. No major clotting difficulties have been observed.

Other modifications of the immunoadsorbents such as heparin-bonding[34,35] may further improve the biocompatibility of the immunoadsorbents.

SUMMARY

The technology of selective removal of anti-red cell antibodies is an exciting achievement. Patients undergoing ABO-incompatible marrow transplant can now have a treatment that specifically removes anti-A or anti-B red cell antibodies prior to marrow infusion. This technique should allow solid organ transplantation from ABO-incompatible donors. Preliminary studies suggest that other red cell blood group antigens (Lewis) could theoretically be adsorbed and possibly improve kidney graft survival in certain situations. Most importantly, these studies demonstrate very clearly that as new antigens are synthesized, the technology is available to utilize them clinically. Our goal is to be able selectively to remove HLA and platelet antibodies in the same manner as A and B antibody. Finally, the refinement of immunoadsorption for use with whole blood will decrease cost and increase utility of this technique.

REFERENCES

1. Landsteiner K, Witt DH: Observations on the human blood groups: Irregular reactions; isoagglutinins in sera of group IV; the factor A. *J Immunol* 1926; 11:221.

2. Hakomori S-I: Blood group ABH and Ii antigens of human erythrocytes: Chemistry, polymorphism, and their developmental change. *Semin Hematol* 1981; 18:39.

3. Wiener AS: Origin of naturally occurring hemagglutinins and hemolysins: A review. *J Immunol* 1951; 66:287.

4. Duquesnoy RJ, Anderson AJ, Tomasulo PA, et al: ABO compatibility and platelet transfusions of alloimmunized thrombocytopenic patients. *Blood* 1979; 54:595.

5. Jaffe EA, Nachman RL, Becker CG, et al: Culture of human endothelial cells derived from umbilical veins: Identification by morphologic and immunologic criteria. *J Clin Invest* 1973; 52:2745.

6. Paul LC, van Es LA, de la Riviere GB, et al: Blood group B antigen on renal endothelium as the target for rejection in an ABO-incompatible recipient. *Transplantation* 1978; 26:268.

7. Thomas ED, Lochte HL Jr, Cannon JH, et al: Supralethal whole body irradiation and isologous marrow transplantation in man. *J Clin Invest* 1959; 38:1709.

8. Hume DM, Merrill JP, Miller BF, et al: Experiences with renal homotransplantation in the human: Report of nine cases. *J Clin Invest* 1955; 34:327.

9. Starzl TE, Marchioro TL, Waddell WR: Human renal homotransplantation in the presence of blood group incompatibilities. *Proc Soc Exp Biol Med* 1963, 113:471.

10. Starzl TE, Marchioro TL, Holmes JH, et al: The incidence, cause, and significance of immediate and delayed oliguria or anuria after human renal transplantation. *Surg Gynecol Obstet* 1964; 118:819.

11. Hamburger J, Crosnier J, Dormont J: Experience with 45 renal homotransplantations in man. *Lancet* 1965; 1:985.

12. Bleyer WA, Blaese RM, Bujak JS, et al: Long-term remission from acute myelogenous leukemia after bone marrow transplantation and recovery from acute graft-versus-host reaction and prolonged immunocompetence. *Blood* 1975; 45:171.

13. Karhi KK, Andersson LC, Vuopio P, et al: Expression of blood group A antigens in human bone marrow cells. *Blood* 1981; 57:147.

14. Gale RP, Feig S, Ho W, et al: ABO blood group system and bone marrow transplantation. *Blood* 1977; 50:185.

15. Buckner CD, Clift RA, Sanders JE, et al: ABO-incompatible marrow transplants. *Transplantation* 1978; 26:233.

16. Bensinger WI, Buckner CD, Thomas ED, et al: ABO-incompatible marrow transplants. *Transplantation* 1982; 33:427.

17. Buskard NA, Williams B, Terman DS, et al: Antibody adsorption on formalinized red cell ghosts: A potential method for the in vivo removal of isohemagglutinins. *Blood* 1978; 52 (Suppl 1):296 (abstract).

18. Pirofsky B, Rosner ER: DTT test: a new method to differentiate IgM and IgG erythrocyte antibodies. *Vox Sang* 1974; 27:480.

19. Lemieux RU: Haworth Memorial Lecture: Human blood groups and carbohydrate chemistry. *Chem Soc Rev* 1978; 7:423

20. Mollison PL: *Blood Transfusion in Clinical Medicine*. Fifth edition. Oxford, Blackwell Scientific Publications, 1972.

21. Bensinger WI, Baker DA, Buckner CD, et al: In vitro and in vivo removal of anti-A erythrocyte antibody by adsorption to a synthetic immunoadsorbent. *Transfusion* 1981; 21:335.

22. Minden P, Farr RS: The ammonium sulphate method to measure antigen-binding capacity. In *Handbook of Experimental Immunology*. Edited by DM Weir. Oxford, Blackwell Scientific Publications, 1967, p 463.

23. Bensinger WI, Baker DA, Buckner CD, et al: Immunoadsorption for removal of A and B blood-group antibodies. *N Engl J Med* 1981; 304:160.

24. Bensinger WI: Plasma exchange and immunoadsorption for removal of antibodies prior to ABO incompatible bone marrow transplant. *Artif Organs* 1981; 5:254.

25. Braine HG, Sensenbrenner LL, Wright SK, et al: Bone marrow transplantation with major ABO blood group incompatibility using erythrocyte depletion of marrow prior to infusion. *Blood* 1982; 60:420.

26. Storb R, Prentice RL, Thomas ED: Marrow transplantation for treatment of aplastic anemia: An analysis of factors associated with graft rejection. *N Engl J Med* 1977; 296:61.

27. Porter KA: Morphological aspects of renal homograft rejection. *Br Med Bull* 1965; 21:171.

28. Slapak M, Naik RB, Lee HA: Renal transplant in a patient with major donor-recipient blood group incompatibility: Reversal of acute rejection by the use of modified plasmapheresis. *Transplantation* 1981; 31:4.

29. Brynger H, Blohmé I, Lindholm A, et al: Transplantation of cadaver kidneys from blood group A_2 donors. *Transplant Proc* 1982; 14:195.

30. Wilbrandt R, Tung KSK, Deodhar SD, et al: ABO blood group incompatibility in human renal homotransplantation. *Am J Clin Pathol* 1969; 51:15.

31. Oriol R, Cartron J, Yvart J, et al: The Lewis system: New histocompatibility antigens in renal transplantation. *Lancet* 1978; 1:574.

32. Williams G, Pegrum GD, Evans CA: Lewis antigens in renal transplantation. *Lancet* 1978; 1:878 (letter to the editor).

33. Bensinger WI, Buckner CD, Baker DA, et al: Removal of specific antibody in vivo by whole blood immunoadsorption: Preliminary results in dogs. *J Clin Apher* 1982; 1:2.

34. Chang TMS: Platelet-surface interaction: Effect of albumin coating or heparin complexing on thrombogenic surfaces. *Can J Physiol Pharmacol* 1974; 52:275.

35. Mangano DT: Heparin bonding and long-term protection against thrombogenesis. *N Engl J Med* 1982; 307:894 (letter to the editor).

CHAPTER 4

The Removal of Amino Acids from Plasma

Gottfried Schmer, M.D.

INTRODUCTION

This chapter deals with the use of immobilized enzymes for the removal of amino acids from plasma. *Escherichia coli* asparaginase (E.C. 3.5.1.1), *Acinetobacter* glutamin-(asparagin)-ase (E.C. 3.5.1.38), indolyl-3-alkane α-hydroxylase (a tryptophan-degrading enzyme), tyrosinase (E.C. 1.14.18.1) and phenylalanine ammonia-lyase (E.C. 4.3.1.5) will be discussed. L-Asparagine, L-glutamine, and L-tryptophan are recognized today as important nutrients for cancer cells, and their removal from the cancers' environment can induce remissions in tumors.[1-3] Tyrosine concentration in plasma is increased in fulminant hepatic coma and, as a precursor of false neurotransmitters, it seems to play an important role in the pathogenesis of hepatic coma.[4] An extensive review of the use of amino acid-depleting enzymes in cancer therapy was published recently.[5]

Because parenteral administration of the enzymes listed above, which are derived from bacterial and fungal sources, can lead to immunological responses,[6] it seems logical to bind these enzymes to water-insoluble polymers (creating a so-called bioreactor) suitable for blood or plasma perfusion.

PROBLEMS ASSOCIATED WITH ENZYME REACTORS (BIOREACTORS)

Activity

Enzyme reactor activity under saturated substrate conditions (V_{max} [maximum velocity]) is the least of the problems. Activity generally has been sufficient for the different enzyme reactor systems described.

Substrate Affinity

A more serious problem is the frequently observed change in the affinity constant (K_m app) of enzymes bound to water-insoluble matrices.[7] L-Asparaginase coupled to a Dacron vascular prosthesis subjected to a flow of 100 back-and-forth strokes per minute showed an increase in K_m app—from 1.2×10^{-5} M for free asparaginase to 2.0×10^{-3} M in its insolubilized form—reflecting a considerable loss of substrate affinity.[8] The same increase in K_m app was found in nylon tubing (internal diameter, 0.1 mm) containing covalently bound L-asparaginase: K_m app increased to 3×10^{-3} M with a substrate flow rate of 1.8 ml/min and to 6.7×10^{-4} M at a flow rate of 4.3 ml/min.[9] In contrast to these results, L-asparaginase coupled to nylon tubing via a polyanionic surface, rather than to a glutaraldehyde linkage as in the above report, gave a K_m app under comparable flow conditions of 2.1×10^{-5} M, approaching the K_m of free asparaginase.[10] Another report[11] dealing with L-asparaginase coupled to Dacron fibers by means of γ-aminopropyltriethoxysilane and glutaraldehyde also described a strong dependence of K_m

app on flow rate: at flow rates of 250 ml/min through this enzyme reactor unit, a K_m approaching that of free asparaginase was found.

These results stress two important points—namely, that differences in the chemistry of bonding lead to different values for K_m app at comparable flow rates as a function of a different "microenvironment" around the enzyme and that the K_m app is diffusion controlled. The practical implication therefore is that extracorporeal enzyme reactor systems should be used at high blood or plasma perfusion rates.

Stability

Stability, in vitro and particularly in vivo, is an important problem that has not yet been solved ideally. The cyanogen bromide activation method, once widely used for the coupling of enzymes,[12] has been replaced in popularity by cyanuric chloride activation[13] which yields a stable, nonhydrolyzable bond. Because the cyanogen bromide method yields isourea bonds that are labile to hydrolysis, enzyme leakage can be appreciable,[14] particularly in the presence of strong nucleophiles.[15] Even bonds created by the γ-aminopropyltriethoxysilane-glutaraldehyde activation method and thought to be fairly stable show some instability.[11]

The in vivo stability of matrix-bound enzymes exposed to blood or plasma represents a formidable problem. Asparaginase bonded to Dacron and implanted into dogs showed an appreciable loss of enzyme activity within the experimental period of 10 days.[8] An enzyme leakage of 7×10^{-4} unit/ml has been observed in dogs when L-asparaginase coupled to Dacron wool was enclosed in an enzyme reactor system exposed to extracorporeal circulation.[11] Although this amount is small, it has to be kept in mind that a leakage of an anti-tumor enzyme like indolyl-3-alkane α-hydroxylase with a K_m of 1.0×10^{-6} M could lead to a false assessment of the amino acid depletion kinetics in plasma produced by the insolubilized enzyme. It follows, therefore, that any kinetic evaluation of amino acid depletion in plasma achieved by extracorporeal enzyme therapy must be corrected for the exact degree of enzyme leakage into the bloodstream.

It now is thought that a major problem of enzyme inactivation in enzyme reactor systems exposed to extracorporeal flow of blood or plasma is digestion of the ligand by plasmatic enzymes, which releases inactive fragments into the general blood circulation and causes immunoallergic reactions.[16] The possibilities of triggering the activation of enzyme systems that cause such phenomena as fibrinolysis, complement reactions, and platelet release reactions during extracorporeal blood circulation are numerous. Blockage of the amino groups of insolubilized asparaginase by succinylation proved successful in increasing the in vivo resistance of the enzyme reactor in sheep.[16]

Sterilization

Considerable progress has been made in the sterilization of enzyme reactors. Hollow fiber enzyme reactors have been sterilized with an ethylene oxide-Freon mixture. Little inactivation of the matrix-bound arginase occurred.[17] Sterilization of enzyme reactors with sodium azide is carried out routinely in our laboratory and has proved to be safe in more than 50 animal experiments.[16]

Choice of Animal Model

The importance of the specific animal model in the kinetics of amino acid depletion by extracorporeal enzyme therapy has become evident, particularly in dealing with the phenomenon of amino acid rebound in plasma. In sheep, the removal of L-asparagine and L-glutamine by an extracorporeal *E. coli* asparaginase reactor system was followed by a protracted decrease in both amino acids for 20 hours after the experiment.[16] The decrease in L-glutamine was caused by the small amount (2—5%) of glutaminase activity of *E. coli* asparaginase. When the same reactor system was used in dogs, there was a rapid increase in L-asparagine and L-glutamine as soon as the extracorporeal enzyme treatment was stopped.[18] These results are in agreement with published rebound kinetics for asparagine in dogs[11] and in man and rats.[19] Cells have an enormous capacity to resynthesize L-glutamine, and overcoming the fast rebound of this important amino acid after its removal from plasma will be a formidable challenge in the field of extracorporeal enzyme therapy. A long-lasting depression of L-glutamine levels in plasma by extracorporeal enzyme reactors would greatly expand the field of cancer therapy by amino acid-degrading enzymes because L-glutaminase has a much broader specificity than L-asparaginase but it also has a much higher toxicity.[5]

REMOVAL OF AMINO ACIDS FROM PLASMA IN VIVO BY EXTRACORPOREAL ENZYME TREATMENT

The V_{max} and the substrate affinity of immobilized enzymes, and therefore their efficiency in depleting plasma of specific components, depend greatly on the microenvironment of the bound enzymes, which depends on the physicochemical characteristics of the supporting matrix. It appears appropriate to divide the complex topic of extracorporeal enzyme treatment for the removal of amino acids into categories according to the underlying supporting matrix used.

Artificial Cells

The synthesis of "artificial cells"—semipermeable aqueous microcapsules—represents a major development in the field,[20] especially for use in an extracorporeal shunt system.

The high activity and the stability of the microencapsulated enzymes are impressive, but the system suffers from the drawback of not being applicable to plasma solutes with strong binding affinities to proteins. A nice example of the application of this system is the immobilization of tyrosinase within collodion microcapsules for potential detoxification in liver failure.[21] This system has been used in rats with galactosamine-induced fulminant hepatic failure because tyrosinase not only converts tyrosine, a precursor of false neurotransmitters, to *o*-quinone but also oxidizes the highly toxic phenol that is present in an increased amount in plasma during hepatic coma.[22] A decrease in tyrosine concentration from 14.44 ± 3.72 mg/dl (mean $\pm$ SD; $n = 9$ galactosamine-treated rats) to 6.83 ± 2.03 mg/dl was observed after 1 hour of hemoperfusion. At the same time, the total free phenol concentration decreased from 0.49 ± 0.12 mmol/liter to 0.09 ± 0.085 mmol/liter.

Another kind of artificial cell is the erythrocyte ghost containing entrapped enzymes. The method of entrapping enzymes within ghost membranes is simple and efficient[23] and the vast literature has been recently reviewed.[5] These artificial cells can be used in an extracorporeal system similar to the collodion capsule one[21] or after the enzyme-filled erythrocyte ghost is bound to a water-insoluble matrix.[24] The same restrictions apply as with semipermeable collodion membranes, and additional factors decrease the overall use of the system to a few specific examples. The selectivity of the erythrocyte membrane prevents some amino acids—e.g., L-asparagine and L-glutamine—from diffusing freely through the membrane, which causes an increase in the observed K_m app of entrapped L-glutaminase and L-asparaginase.[24]

Two other examples, not strictly belonging to the field of extracorporeal amino acid depletion, are mentioned here because the extracorporeal method is applicable. They involve intravenous injection of artificial cells (erythrocyte membrane entrapment) into experimental animals.

In one, L-asparaginase entrapped in erythrocyte ghosts was injected into monkeys.[25] There was total depletion of L-asparagine for 20 days after a single injection. In contrast, L-asparagine could be depleted for only 10 days after a comparable injection of free enzyme. The authors stressed the advantage of this form of enzyme therapy, but they explained the depletion of L-asparagine from plasma on the basis of a slow release of enzyme from the ghosts. This interpretation makes sense in the light of later findings.[24]

In the other example, depletion of L-tryptophan in the plasma of mice was observed after the injection of ghosts filled with indolyl-3-alkane α-hydroxylase.[25] The results were similar to those found for entrapped L-asparaginase,[26] namely, about a twofold prolongation of enzyme activity and a total depletion of L-tryptophan in plasma, from 100 nmol/ml to less than 5 nmol/ml. Although no free enzyme activity could be found in plasma with the methods used, it seems also likely that the depletion of L-tryptophan was caused by the release of minute amounts of enzymes because L-tryptophan is strongly protein bound and cannot readily diffuse through the membrane of the erythrocyte ghost.

Matrices Derived from Biological Materials

The present state of this field has been recently reviewed.[27] The great advantage of using biological materials as a support for enzymes is the generally very high ratio of enzyme activity to support weight. An obvious disadvantage is the mechanical vulnerability of the matrix. L-Asparaginase bound to implantable collagen heterografts was used in dogs as a potential long-term sustained-dosage antitumor enzyme therapy system.[28] The collagen was from de-endothelialized bovine arteries and the enzyme was bound by glutaraldehyde cross-linkage. Four normal dogs received an arterial implant, and three showed a prolonged depression of L-asparagine levels in the plasma, from a mean of 37 nmol/ml to 10 nmol/ml. Clotting on the grafts, development of a fibrous layer on the inner surface, considerable resorption, and a loss of L-asparaginase activity—64% after 4½ days with one graft and 70% after 13 days with another graft—represent the problems associated with this method. In spite of its failure, this work deserves consideration because it addresses two crucial points in the field of enzyme therapy by use of matrix-bound enzymes: (1) exposure of immobilized enzymes to very high blood flows, which decreases the K_m app appreciably by decreasing the diffusion layer for the substrate,[8–10] and (2) counteracting the rebound of amino acids after temporary extracorporeal enzyme therapy by using an implant for prolonged suppression of amino acid levels.

Collagen membranes have been used as a matrix for insolubilized L-asparaginase in an extracorporeal enzyme reactor system in healthy dogs.[29] The high enzyme activity, 250 units/ g of membrane, reflects the well-known excellent enzyme-to-support ratio of bioartificial organs. Hemoperfusions were run for 1–2 hours in eight dogs and decreased the L-asparagine levels in plasma to less than 0.5 nmol/ml from starting values of 20–35 nmol/ml. The depletion occurred as early as 15 minutes after start of perfusion. This appears to be somewhat too precipitous, and one must question whether an initial leakage of L-asparaginase into the blood may have caused an analytical artifact as discussed in regard to difficulties of assessing extracorporeal enzyme reactor efficiency.

In one experiment, after 1 hour of hemoperfusion the L-asparagine level stayed down for 24 hours, which is contrary to the general experience of rebound of L-asparagine in dogs. However, the experiments clearly showed the considerable mechanical and chemical stability of the L-asparaginase-collagen membrane. In 12 trials with in vivo contact of 12 hours or more, no loss of enzyme activity could be found. These positive characteristics together with the high specific activity of the enzyme-matrix complex should encourage more work in this special area.

The idea of using biological material as an insoluble support for L-asparaginase has been extended to polymerized fibrin.[27,30] Fibrin cast on roughened glass plates (10 × 10 cm) was treated with crude factor XIII and a 10% (vol/vol) glutaraldehyde solution which not only further cross-linked the fibrin polymer but also activated the amino groups for the attachment of L-asparaginase. This cross-linked fibrin polymer was completely stable against fibrinolytic attack and degraded 2,800 μmol of L-asparagine per min per m^2 of polymer surface at 37°C. When a parallel plate-organ consisting of L-asparaginase-fibrin plates was used as an extracorporeal enzyme reactor device in healthy sheep, L-asparagine levels dropped from a mean of 53 nmol/ml to 15 nmol/ml plasma after 6 hours of extracorporeal enzyme treatment. No leakage of L-asparaginase into blood was observed on analysis with a sensitive radioassay that is able to detect as little as 10^{-5} units of L-asparaginase per ml of plasma. No loss of enzyme activity in the fibrin plate-organ could be detected after 18 hours of in vivo exposure.

The idea of an L-asparaginase implant in the form of a vascular prosthesis or a bovine arterial implant[8,28] has been pursued by using micropore (0.55 μm) polypropylene tubing with fibrin gel in the porous wall. L-Asparaginase was coupled to the fibrin mesh as described above and used ex vivo as a connecting tubing in a carotid-jugular shunt system.[31] In three preliminary experiments of 24−48 hours' duration, in sheep there was a decrease of the L-asparagine level to 25−50% of its baseline values with a residual enzyme activity in the microreactor tubing of 70% after removal from the shunt system. Total L-asparaginase activity for a 30-cm-long piece of tubing was 35 units at 37°C, which is in agreement with the high enzyme activity found in the L-asparaginase-fibrin plate-organ. It is conceivable that the combination of an extracorporeal enzyme reactor system with this type of microreactor tubing applied ex vivo between the main hemoperfusion treatments could improve the efficiency of the treatment considerably by preventing the fast rebound of L-asparagine.

Hollow-Fiber Artificial Kidneys (HFAK)

Advantages of the use of an HFAK device[13] as the basis for an enzyme reactor system include its commercial availability at reasonable cost, the possibility of high blood flow rates, its mechanical stability, and the possibility of activating the cellulose fibers with a wide variety of chemical substances for the covalent binding of amino acid-depleting enzymes.

L-Asparaginase has been covalently bound to the outer surface (dialysis side) of the fibers by use of various binding techniques (cyanogen bromide, cyanuric chloride, and glutaraldehyde.[32] The glutaraldehyde-cross-linked L-asparaginase reactor showed a high degree of stability in vitro. The cyanogen bromide-activated enzyme reactor was less stable, presumably because of hydrolysis of the enzyme-matrix bond. The advantage of binding the enzyme on the outer surface of the fibers is the prevention of an immune response that might be provoked by a leakage of enzymes into the bloodstream.

On the other hand, this system has severe limitations and disadvantages, such as its widely decreased usefulness for degrading amino acids—e.g., tryptophan—strongly bound to protein and its high K_m app, which reflects low apparent substrate affinities. The K_m app for L-asparaginase in this system is 1.0×10^{-3} M, about 100 times higher than the K_m of the free enzyme. This reflects the dependence of the system on the rate of solute transfer to the membrane. Although the mechanical characteristics of the HFAK remained largely unchanged after the bonding of L-asparaginase, some decrease in permeability through the membrane was observed.

This HFAK-enzyme reactor system was tested in three patients, one with acute lymphoblastic leukemia, one with poorly differentiated lymphocytic lymphoma, and one with acute rejection episodes after transplantation of a cadaver kidney.[13] In both tumor patients, L-asparagine in plasma decreased to undetectable levels after 3 hours of extracorporeal enzyme therapy. One patient, who had had allergic reactions upon the parenteral administration of L-asparaginase previously, tolerated the extracorporeal enzyme therapy without any immunological reactions. Both tumor patients showed a transient remission. The patient with lymphoblastic leukemia did not seem to respond to the therapy after 1 week, with L-asparagine levels increasing to 60—120 nmol/ml and a successive increase in blast counts. The patient with lymphoma showed a blast crisis after initial remission in spite of low L-asparagine levels in plasma, indicating tumor resistance. Extracorporeal L-asparaginase treatment had no effect in the patient with a transplant rejection crisis.

It is noteworthy that, in spite of all the disadvantages described above for HFAK-enzyme reactors containing the enzyme on the dialysis side and consequently having low substrate affinities, total depletion of L-asparagine in plasma could be obtained with blood flows of 200 ml/min. It also demonstrates from a practical standpoint the feasibility of a routine HFAK-enzyme reactor program in a dialysis unit, combining the overall simple enzyme binding method with the clinical expertise for routine extracorporeal therapy.

The disadvantage of using extracorporeal enzyme treatment with immobilized L-asparaginase alone is based on the relatively fast rebound of L-asparagine in plasma after depletion because of the antagonistic action of L-asparagine synthetase. In higher animals and in man, L-glutamine serves as the ammonia donor for the synthesis of L-asparagine. A strategy was therefore developed to decrease L-asparagine and L-glutamine simultaneously in plasma to prevent the rebound of L-asparagine.[19] A mini HFAK-enzyme reactor[32] containing bound L-asparaginase alone at the outer side of the filter was used in three normal rats. After a decrease in mean L-asparagine levels in plasma from 1.04 ± 0.13 mg/dl to 0.03 ± 0.01 mg/dl by 2 hours of treatment, a normal level, 0.92 ± 0.1 mg/dl, could be found after 24 hours. Two normal rats were treated with a combined L-asparaginase/L-glutaminase reactor in which L-glutaminase was slowly circulated through a closed circuit on the outside of the L-asparaginase reactor described above. L-Asparagine fell to undetectable levels after 2 hours of extracorporeal treatment; L-glutamine decreased to about 50% of its baseline values. At 24 hours later, L-asparagine levels remained depressed to about 30% of their baseline values and L-glutamine still was about 60% of pretreatment values. The importance of these experiments is the delay in the L-asparagine rebound after extracorporeal enzyme therapy, achieved by using a combined L-asparaginase/L-glutaminase reactor system. The concept of an extracorporeal L-asparaginase/L-glutaminase reactor greatly broadens the applicability of extracorporeal enzyme treatment of tumors sensitive to L-asparagine and L-glutamine deprivation.

To overcome the disadvantages of HFAK-enzyme reactors with enzymes bound to the outside of the fibers, some studies have been performed with beef liver arginase covalently bound on the inside of the hollow fibers.[17] Although several tumors that are sensitive to arginine deprivation have been described, it appears that this enzyme is of minor importance as an antitumor enzyme because of it high K_m. Immobilization of arginase on the cellulose fibers was carried out by using the metaperiodate oxidation method which resulted in the immobilization of $0.3\mu g$ of active enzyme per cm^2 and the generation of 250 μmol of urea per minute, a remarkable activity. The immobilization did not affect adversely the physical and mechanical properties of the hollow fibers. Sterilization with ethylene oxide-Freon caused only a 30% decrease of the original activity. The extracorporeal use in sheep was well tolerated. Blood and liver function values were not different from those of normal hemodialysis. Arginine values before and after extracorporeal enzyme treatment were not given.

An interesting approach is the use of a multitubular hollow-fiber device with phenylalanine ammonia-lyase entrapped in the fiber wall.[33,34] This enzyme converts phenylalanine to *trans*-cinnamic acid. Phenylalanine accumulates in the blood of patients with phenylketonuria (PKU), a hereditary deficiency of phenylalanine hydroxylase. This accumulation leads to mental retardation. Although the disease is well treated by restricting the patient to a phenylalanine-free diet, certain crisis situations occur in these patients, such as with fever or infection during pregnancy, when the genetically normal fetus can suffer brain damage by the mother's high phenylalanine levels. In these situations the possibility of lowering the phenylalanine levels with extracorporeal enzyme therapy has been suggested.

The hollow-fiber enzyme reactor containing phenylalanine ammonia-lyase has been used in dogs in which the phenylalanine level was increased artificially by diet and the administration of p-chlorophenylalanine, a phenylalanine hydroxylase inhibitor. Under extracorporeal enzyme therapy, the increased amino acid levels decreased to 18% of their pretherapy levels and stayed around 50% for 5 days. No leakage of the entrapped enzyme was reported. The K_m app of the insolubilized phenylalanine ammonia-lyase was 5.0×10^{-4} M, which was identical to the K_m of the free enzyme and indicated that no diffusional effects were involved.

Other Systems

Extracorporeal perfusion treatment with immobilized L-asparaginase was carried out by using a parallel plate-organ of poly(methyl methacrylate) matrix.[35,36] Two patients with melanoma were treated with extracorporeal enzyme therapy for 80 and 99 hours, respectively. A decrease of L-asparagine levels in plasma was achieved, from 30 nmol/ml to 7 nmol/ml. Most importantly, no allergic reactions were observed in the patients. The same enzyme reactor system was used in patients with kidney transplants as an adjunctive immunosuppressive therapy; results were disappointing[37] although L-asparagine levels decreased to undetectable levels during the extracorporeal enzyme therapy.

Plasmapheresis-Enzyme Reactor System

All enzyme reactor systems described above have the common disadvantage of direct contact with the cellular elements of blood. In particular, platelets and granulocytes tend to adhere to foreign surfaces, covering them with a dense cell layer and thus decreasing the interaction between the bound enzyme and the substrate. Platelets passing through an artificial organ are also known to show a considerable release of enzymes and specific platelet proteins, such as a heparin-neutralizing factor (platelet factor 4). The enzymes in particular represent a danger to the integrity of the insolubilized protein ligands. The use of an extracorporeal enzyme reactor, therefore, always has an element of unpredictability regarding the final result achieved. In addition, insolubilized enzymes such as aparaginase are in contact with macrophages and B lymphocytes, cellular elements of the immune system.

To avoid these disadvantages, a new enzyme reactor system has been introduced—the plasmapheresis enzyme reactor system.[16] Blood is passed from a carotid-jugular shunt through a polypropylene plasma filter. The plasma is filtered through the membranes at 20 ml/min and then passed through a cartridge containing granular cellulose to which L-asparaginase is attached by a nonhydrolyzable isocyanate bond. (The newer types of filter allow plasma flow rates in excess of 50 ml/minute.) The L-asparaginase is made resistant to tryptic digestion by a simple treatment with succinic anhydride, which blocks the free amino groups. (This has opened the way for long-term treatment without loss of enzyme activity or the appearance of allergic symptoms after in vivo exposure for up to 100 hours in dogs with lymphoma.[38] After passing through the enzyme reactor, the plasma is mixed with the formed elements and returned to the animal. A single pass through the column decreased L-asparagine levels from a mean of 63 nmol/ml to 5 nmol/ml. L-Glutamine decreased from 290 nmol/ml to 158 nmol/ml due to the 2−5% L-glutaminase activity present in *E. coli* asparaginase. No enzyme leakage could be detected at the outlet of the reactor (lower limit of detection, 10^{-3} unit/ml). It is interesting that the L-asparagine and L-glutamine levels stayed depressed for more than 20 hours. In the absence of enzyme leakage, this is contrary to the experience in man, dogs, and rats.

The plasmapheresis enzyme reactor system was tolerated very well and the reproducibility of results showed an amazingly narrow range. This system, therefore, has a good chance of clinical application in the future.

SUMMARY AND FUTURE PROSPECTS

This chapter deals mainly with extracorporeal enzyme therapy, with particular reference to the so-called antitumor enzymes, L-asparaginase and L-glutaminase. Removal of amino acids from plasma is considered mainly from the perspective of removing identified cancer nutrients such as L-asparagine, L-glutamine, and L-tryptophan. The field shows promise for successful therapy of tumors sensitive to amino acid-deprivation in regard to two developments: (1) use of extracorporeal enzyme reactors consisting of two or more specific amino acid-degrading enzymes, and (2) combined use of extracorporeal enzyme therapy and amino acid analogues, which would greatly enhance the antitumor effect.

The parenteral administration of indolyl-3-alkane α-hydroxylase and tryptophan analogues has shown great promise in cancer therapy.[39] It is equally clear, however, that our beginnings in extracorporeal enzyme therapy are relatively modest and that concerted efforts are necessary to expand the technique to many more enzymes and chemotherapy combinations.

REFERENCES

1. Hill JM, Roberts J, Loeb E, et al: L-Asparaginase therapy for leukemia and other malignant neoplasms: Remission in human leukemia. *JAMA* 1967; 202:882.
2. Spiers ASD, Wade HE: Bacterial glutaminase in treatment of acute leukemia. *Br Med J* 1976; 1:1317.
3. Roberts J, Schmid FA, Rosenfeld HJ: Biologic and antineoplastic effects of enzyme-mediated in vivo depletion of L-glutamine, L-tryptophan, and L-histidine. *Cancer Treat Rep* 1979; 63:1045.

4. Fischer JE, Funovics JM, Aguirre A, et al: The role of plasma amino acids in hepatic encephalopathy. *Surgery* 1975; 78:276.

5. Holcenberg JC, Roberts J: *Enzymes as Drugs.* New York, John Wiley & Sons, 1981.

6. Peterson RG, Handschumacher RE, Mitchell MS: Immunological responses to L-asparaginase. *J Clin Invest* 1971; 50:1080.

7. Melrose GJH: Insolubilized enzymes: Biochemical applications of synthetic polymers. *Rev Pure Appl Chem* 1971; 21:83.

8. Cooney DA, Weetall HH, Long E: Biochemical and pharmacologic properties of L-asparaginase bonded to Dacron vascular prostheses. *Biochem Pharmacol* 1975; 24:503.

9. Allison JP, Davidson L, Gutierrez-Hartman A, et al: Insolubilization of L-asparaginase by covalent attachment to nylon tubing. *Biochem Biophys Res Commun* 1972; 47:66.

10. Horvath C, Sardi A, Woods JS: L-asparaginase tubes: Kinetic behavior and application in physiological studies. *J Appl Physiol* 1973; 34:181.

11. Ko RYC, Hersh LS: A Dacron wool packed-bed extracorporeal reactor: A kinetic study of immobilized *Escherichia coli* II L-asparaginase. *J Biomed Mater Res* 1976; 10:249.

12. Porath J, Axén R: Immobilization of enzymes to agar, agarose, and Sephadex supports. *Methods Enzymol* 1976; 44:19.

13. Dumler F, Singh PR, Jackson CE, et al: Extracorporeal enzyme therapy: Use of an L-asparaginase reactor-dialyzer in a clinical setting. *ASAIO J* 1981; 4:70.

14. Lasch J, Koelsch R: Enzyme leakage and multipoint attachment of agarose-bound enzyme preparations. *Eur J Biochem* 1978; 82:181.

15. Schnapp J, Shalitin Y: Immobilization of enzymes by covalent binding to amine supports via cyanogen bromide activation. *Biochem Biophys Res Commun* 1976; 70:8.

16. Schmer G, Rastelli LN, Dennis MB, et al: Successful in vivo experiments in sheep using a new nonhydrolyzable enzyme reactor resistant against tryptic digestion. *Trans Am Soc Artif Intern Organs* 1982; 28:374.

17. Rossi V, Malinverni A, Callegaro L: Immobilization of arginase on hollow fiber hemodialyzer. *Int J Artif Organs* 1981; 4:102.

18. Dennis MB, Schmer G, Rastelli LN, et al: Successful long-term use of a plasmapheresis reactor system in dogs with lymphoma. *Trans Am Soc Artif Intern Organs* 1983; 29:744.

19. Giordano C, Esposito R, Mazzola G, et al: L-Glutaminase and L-asparaginase by extracorporeal route in acute lymphoblastic leukemia therapy. *Int J Artif Organs* 1981; 4:244.

20. Chang TMS: Semipermeable aqueous microcapsules "artificial cells": With emphasis on experiments in an extracorporeal shunt system. *Trans Am Soc Artif Intern Organs* 1966; 12:13.

21. Shu CD, Chang TM: Tyrosinase immobilized within artificial cells for detoxification in liver failure: I. Preparation and in vitro studies. *Int J Artif Organs* 1980; 3:287.

22. Barman TE: *Enzyme Handbook.* Vol 1. New York, Springer-Verlag, 1969, p 226.

23. Bodemann H, Passow H: Factors controlling the resealing of the membrane of human erythrocyte ghosts after hypotonic hemolysis. *J Membr Biol* 1972; 8:1.

24. Schmer G, Krueger CM, Cole JJ: Gel-bound resealed red cell membranes: A new type of semi-artificial organ. *Trans Am Soc Artif Intern Organs* 1977; 23:667.

25. Schmer G, Roberts J: Induction of hypothermia in mice by semi-artificial cells containing indolyl-3-alkane α-hydroxylase. *Trans Am Soc Artif Intern Organs* 1979; 25:39.

26. Updike SJ, Wakamiya RT, Lightfoot EN Jr: Asparaginase entrapped in red blood cells: Action and survival. *Science* 1976; 193:681.

27. Schmer G, Rastelli L, Newman ML, et al: The bio-artificial organ: Review and progress report. *Int J Artif Organs* 1981; 4:96.

28. Jefferies SR, Richards R, Bernath FR: Preliminary studies with L-asparaginase bound to implantable bovine collagen heterografts: A potential long-term, sustained dosage, antitumor enzyme therapy system. *Biomater Med Devices Artif Organs* 1977; 5:337.

29. Olanoff LS, Bernath FR, Venkatasubramanian K, et al: Reduction of canine serum asparagine levels by L-asparaginase immobilized on collagen: A potential form of cancer chemotherapy. *Trans Am Soc Artif Intern Organs* 1975; 21:156.

30. Schmer G, Bisbroek, J, Dennis MB, et al: Successful use of two new reactor types in sheep. *Trans Am Soc Artif Intern Organs* 1980; 26:129.

31. Schmer G, Dennis MB: Selective in vivo experiments in sheep using a new non-hydrolyzable enzyme reactor resistant against tryptic digestion. *Am Soc Artif Intern Organs Abstracts* 1982; 11:82.

32. Mazzola G, Vecchio G: Immobilization and characterization of L-asparaginase on hollow fibers. *Int J Artif Organs* 1980; 3:120.

33. Pedersen H, Horvath C, Ambrus CM: Preparation of immobilized L-phenylalanine ammonia-lyase in tubular form for depletion of L-phenylalanine. *Res Commun Chem Pathol Pharmacol* 1978; 20:559.

34. Mirand E, Guthrie R, Paul T: Phenylalanine depletion for the management of phenylketonuria: Use of enzyme reactors with immobilized enzymes. *Science* 1978; 201:837.

35. Hydén H, Gelin L-E, Larsson S, et al: A new specific chemotherapy: A pilot study with an extracorporeal chamber. *Rev Surg* 1974; 31:305.

36. Lange S, Hydén H: Extracorporeal perfusion treatment with immobilized L-asparaginase. *Scand J Haematol* 1977; 18:244.

37. Tapia LT, Hyden H, Wellner D, et al: Extracorporeal treatment with membrane immobilized L-asparaginase in kidney transplantation. *Trans Am Soc Artif Intern Organs* 1977; 23:443.
38. Schmer G: Unpulished data.
39. Roberts J, Rosenfeld H, Dunn B: Antineoplastic and biological effects of selected tryptophan analogs and indolyl-3-alkane α-hydroxylase (IAH). *Proc Am Assoc Cancer Res Am Soc Clin Oncol* 1981; 22:218 (abstract).

CHAPTER 5

Immunoadsorption in the Therapy of Malignant Disease

Charles A. Bowles, Ph.D.
Gerald L. Messerschmidt, M.D.

INTRODUCTION

The selected removal of various molecules within the liquid phase of patients has been discussed for various disease entities in other chapters. This chapter will examine a new field of therapy for malignant disease, i.e., adsorption and modulation of immunoregulatory substances by protein A.

In 1940, Verwey,[1] in his efforts to serologically classify strains of staphylococci, found that one of five fractions (fraction B) was highly antigenic. Similar attempts by Jensen[2] showed that *Staphylococcus aureus*, Cowan strain I, contained a component, antigen A, which was particularly effective in stimulating an immune response in rabbits. With further studies, pre-immunized animals and all samples of normal human serum strongly reacted with this antigen extract.[2,3] Further investigations demonstrated the antigen to be a protein, and the name protein A was adopted.[4,5] This unique reactivity with all serum samples was defined by Forsgren and Sjöquist in 1966.[6] Their experiments tested mixtures of protein A with normal or myeloma IgG, or isolated fragments of immunoglobins, i.e., Fab Fc. They demonstrated that the protein A reactivity was a "pseudoimmune" reaction with nonspecific binding to the Fc region of IgG. Extensive testing in other mammals, birds, fishes, amphebians, or reptiles suggested that protein A evolved as long as 200 million years ago.[7]

Not all subclasses of IgG react with protein A. Kronvall and Williams[8] demonstrated that protein A will bind IgG-1, -2, and -4, but not IgG-3. IgM[9] and IgA[10] also bind to protein A though not as avidly as IgG. IgD[11] and IgE[12] are not significantly reactive. It is of interest that immunoglobulins (antibodies), when bound with specific antigen to form an immune complex, have increased binding affinity to protein A.[13] This interesting observation will be examined more closely.

RATIONALE FOR THERAPY

Many studies in experimental and human tumor systems indicate that lymphocytes from tumor hosts are sensitized to the tumor and are cytotoxic in in vitro assays. The addition of tumor-bearing host's serum to these in vitro cultures often inhibits this cytotoxicity.[14–18] These findings suggested that a serum factor was interfering with cellular immunity, and thus blocking of antitumor immunity resulted in progressive tumor growth. Initially, this effect was felt to be a result of a "blocking antibody." More recent evidence suggests that antibody complexed with tumor-associated antigens, i.e., immune complexes, play a significant role in blocking immune activity.[19–21] Removal of circulating immune complexes by plasmapheresis or thoracic duct drainage has resulted in decreases in serum-blocking ability and some tumor regressions.[22,23] The data revealed that serum blocking ability was rapidly renewed and that plasmapheresis procedures were required every 48 hours to maintain reduced levels of blocking

factor.[22] Plasmapheresis procedures this frequent are techni-
cally difficult, thus frequency prevents the use of this approach
as a cancer therapy modality.

In 1974, Steel et al.[24] demonstrated that the cellular
immunosuppression mediated by tumor host serum could be
reduced by exposure of the serum to *S. aureus*, Cowan strain I.
They hypothesized that protein A removed suppressive factors through binding IgG complexes at the Fc receptor site of
protein A. This study led Bansal et al.[25] to treat a metastatic
colon cancer patient with extracorporeal plasma immunoadsorption using *S. aureus*, Cowan strain I. Plasma separated
from heparinized whole blood by continuous flow plasmapheresis was passed through a 0.2 μm membrane filter containing 30–75 g of *S. aureus* and finally through a charcoal
hemodetoxifier before returning to the patient. A decrease in
tumor size was reported following 20 procedures performed
over a 5-month period. Microcytotoxicity tests performed to
measure levels of blocking factor activity revealed decreases
in these factors after therapy. Blocking remained undetectable
with repeated treatments but returned within 90 hours after
the last treatment, indicating that these blocking factors are
rapidly renewed.[25]

Immunologically, serum IgG levels decreased but no decreases were noted in other nonimmunoglobulin proteins. T
lymphocytes decreased initially and remained depressed during repeated perfusion treatments. These cells returned to
near normal levels if treatments were discontinued for 7–10
days. B lymphocyte levels increased after the procedures.[25]
The authors perfused 1–2 liters of the patient's plasma in
each procedure. This would suggest that a significant amount
of the immune complexes remained in the plasma after each
procedure. It is of interest to note also that an abrupt rise in
immune complexes was observed in the blood in the immediate postperfusion periods, and they postulated that these
immune complexes were of the large size which were incapable of blocking cellular immunity in vitro. It was not clear from
this report what caused this abrupt increase in immune com-

plexes but it does not appear to be through release of *S. aureus* from the filter.

Terman et al.[26] expanded on Bansal's report by treating 12 dogs with mammary carcinoma using a similar extracorporeal plasma perfusion procedure. Plasma from tumor-bearing dogs was passed over either *S. aureus*, Cowan strain I (1 kg body weight) or *S. aureus*, Wood strain. The bacteria were immobilized on a 0.2 μm membrane filter system. The Wood strain of *S. aureus*, deficient in protein A, was used to determine the role of this protein in eliciting tumor responses. No tumor responses were reported following plasma perfusion over the Wood strain bacteria while seven of 12 dogs treated with Cowan strain I had objective (greater than 50%) decreases in measurable tumor. These authors reported microscopic tumor necrosis beginning 4 hours after perfusion.[26] Six of six responding dogs had necrosis of ulcerated lesions with healing of the ulcerated areas as early as day 8 after perfusion. They reported immunohistochemical deposits of IgG and C3, and ultrastructural evidence of lytic lesions on tumor cell membranes.[26] Serum IgG and C3 levels decreased and immune complexes rose after perfusion and remained elevated for 72 hours. They interpreted these data to indicate that the acute tumoricidal response was initiated by complement-dependent tumor-specific antibodies freshly generated or rendered operational after extracorporeal plasma perfusion over *S. aureus*, Cowan strain I. The authors speculate that the role of protein A in these studies was to initiate the tumor-specific antibody activation by removal of humoral factors or by dissociation of tumor-specific antibody from substances interfering with functional activity.

Tumor responses following treatment with *S. aureus*, Cowan strain I, have also been observed in dogs with mammary carcinoma in investigations funded by the NCI.[27,28] In our studies, mammary tumors on the chest wall of dogs served as the index lesion to reflect tumor responses. The plasma of 15 randomized dogs was separated by continuous flow centrifugation and passed through pleated membrane filters containing either 0.2 g/kg body weight of Cowan strain I *S. aureus,* (10 dogs) or filters free of *S. aureus* (five dogs). Passage of 50% to 100% of the dogs' plasma over blank filters resulted in no reductions in tumor size in any of the five dogs and no toxicity was observed. Five of the 10 dogs whose plasma was exposed to chambers containing *S. aureus* had objective responses of greater than 50% tumor shrinkage.[27,28] Two of these dogs had tumor regressions to 16% and 10% of the original tumor size. After surgical excision of the remaining tumor lesion, these dogs experienced tumor-free survival of greater than 828 days. The three other responding animals had survival times of 18 to 492 days (mean 460.4 days) while the five nonresponding dogs had survival of 63 to 494 days (mean 245.2 days).

To examine the question of nonspecific substances eluting off the bacteria which could elicit these tumor response effects, we selected five additional dogs with mammary carcinoma for plasma infusion treatments.[28] In these studies, units of whole blood were obtained from normal donors and the plasma was separated by centrifugation. The separated plasma was passed over identical *S. aureus* columns and subsequently infused into the tumor-bearing dogs. Each tumor dog received the equivalent of 20% of its calculated plasma volume once weekly for 4 weeks. No tumor responses were observed over a 4-month observation period.[28]

Analysis of the variables related to tumor response in *S. aureus* perfused dogs showed that the amount of circulating immune complexes prior to treatment was not critical between responding (mean 3.4 immune complex types) and nonresponding (mean 4.6 immune complex types) dogs.[27] In fact, a similar decrease in immune complex levels was observed in the responder (35.3% reduced) and nonresponder (34.9% reduced) animals. Of importance was the fact that nonresponders had significantly more tumor bulk at the initiation of therapy (mean 57.6 cm nonresponders versus mean 23.8 cm^2 for responders), and had more plasma perfusion procedures (mean 3.6 procedures for nonresponders) than responders (mean 2.6 procedure). The variables of age, weight, carcinoma cell type, and previous medical history had no influence on the subsequent response to perfusion. These observations suggested that removal of immunosuppressive factors was not responsible for the tumor response. Perhaps a factor from the adsorbent chamber, inherent or generated, was responsible for the observed responses although our initial studies did not support this concept.[28]

The interpretation that direct removal of blocking or suppressive factors by *S. aureus* immunoadsorption is not the primary mechanism of action for tumor response following perfusion was supported by Jones et al.[29] These authors treated feline leukemia virus (FeLV) infected cats with ex vivo immunoadsorption using 20−30 g/kg body weight of *S. aureus*, Cowan strain I. The cats' plasma was immediately infused after *S. aureus* treatment. Prior to twice weekly immunoadsorption procedures, the cats were subjected to 250−300 rads of whole body irradiation. The combined regimen of irradiation and immunoadsorption resulted in a reversal of FeLV status in five of five cats. Two of these animals developed complete tumor-free status which lasted 7 to 8 months after therapy. Although interpretation of these findings are difficult because of the combined treatment of perfusion and irradiation, these authors postulated that the observed effects resulted from upsetting a static immune balance by perfusion of a small amount of plasma (25% of the plasma volume) rather than from removal of a significant quantity of immune complexes or blocking factors. Consideration must also be given to the fact that tumor lymphoblasts infected with FeLV have antigenic characteristics different from spontaneous tumor cells, and alterations in immune activity by perfusion and irradiation may be different from that directed against virus-free tumor cells, i.e., spontaneous canine mammary carcinoma. Nevertheless, the data supports a role for protein A as an active modulator of tumor responses.

In all of the studies mentioned, immunoadsorption was performed with formalin-fixed, heat-killed *S. aureus* immobilized on a membrane filter. Taken together, the data supported a role for protein A as the active molecule that elicited tumor responses. To document this, Terman et al.[30] perfused plasma from dogs with mammary adenocarcinoma over a preparation of purified protein A (1 mg/kg body weight) bound passively to charcoal by collodion. In these studies, five tumor-bearing dogs received a single infusion of cytosine arabinoside (10 mg/kg body weight) immediately after extracorporeal perfusion of one plasma volume. The authors reported that this treatment regimen resulted in a rapid, diffuse necrosis in the tumors of all dogs. Up to 50% of the surface area of the visible lesions healed in 12 to 18 days after treatment. Regrowth of the tumor occurred in 25 to 37 days. The use of a chemotherapeutic agent in this treatment regimen obfuscates the tumor response effects of protein A; however, the data supported the interpretation that protein A was the active molecule of *S. aureus* which modulated tumor responses.

HUMAN TRIALS

Human studies have now been performed using both *S. aureus* and purified protein A. We have treated five patients in phase I trials with *S. aureus* immunoperfusion similar to the Bansal therapy.[31] No tumor responses were seen in any of the tumors treated which included synovial cell carcinoma, esophageal carcinoma, mammary carcinoma, colon carcinoma, and melanoma. This lack of response may be secondary to small amounts of plasma treated, i.e., 24−60% of the total calculated plasma volume. In our animal studies, tumor responses were observed only in those dogs treated with the largest amount of perfused plasma.[27] Patients, however, experienced toxic reactions during perfusion, thus forbidding continued plasma treatment. Initially, cardiac index increased and resulted in an elevation of the blood pressure. After approximately 30−60 minutes the blood pressure began to fall. Fluids (8/8 procedures) and vasoconstrictors (6/8 procedures) were required to maintain the mean arterial blood pressure above 60 mm Hg.[31] The cardiac index remained elevated. The fall in blood pressure appeared to be secondary to a dramatic decrease in systemic vascular resistance. Concomitant with these changes, respiratory rates increased in all patients. The pCO_2 fell, reflecting the increase in ventilation, and oxygen exchange decreased. Two patients died from inadequate respiratory function.[31]

Terman et al.[32] have reported the use of protein A collodion-charcoal immunoperfusion in combination with chemotherapeutic agents in five patients. In those patients treated with large amounts of protein A and high plasma flow rates, similar toxicity was observed as seen in our *S. aureus* trial. Three of Terman's patients had a partial tumor remission following treatment, although the effect of concomitantly used chemotherapeutic agents on these tumor responses is unclear.[32]

We designed a trial based on the work of Terman et al.[32] to evaluate the effects on tumor response of plasma perfusion over protein A immobilized on charcoal by collodion. Plasma collected from the patient by continuous flow plasmapheresis was stored frozen for subsequent passage over protein A cartridges.[33,34] Treated plasma was immediately returned to the patients after passage over protein A. Volumes of plasma treated and infused increased from 50 to 450 ml per procedure. In eight patients treated, no toxicity has been observed related to the cardiovascular system or the respiratory system. Most procedures have been associated with a fall in platelet counts, and when the procedure was performed daily for 3 days, the platelet toxicity was cumulative.[34] Of importance was that no tumor responses were observed in any patient.

Finally, Bensinger et al.[35] employed immunoadsorbent columns consisting of 200 mg of purified protein A covalently linked to silica. Five patients with breast adenocarcinoma were treated five to ten times by perfusion of cellulose membrane separated plasma. Perfusion procedures performed once or twice per week with 3 liters of plasma produced toxic reactions similar to that observed by others.[25,32,33] Decreases in tumor size to 25 to 50% was observed in three of these patients. These tumor responses were observed in the absence of concomitantly administered chemotherapy as used by Terman et al.[32] Tumor biopsies revealed deposits of IgG and C3 on the cancer cells, similar to those reported found on tumor cells of canine mammary carcinoma.[26] These results further document the therapeutic effectiveness of protein A in ex vivo plasma perfusion.[25-32] Bensinger et al.[35] speculate that the benefits of treatment by protein A immunoadsorption may be related to the binding of IgG or IgG complexes to the immunoadsorbent.

PROTEIN A INFUSION STUDY

Studies relating to plasma perfusion over *S. aureus*, or purified protein A, have documented that significant tumor regressions, leading in some cases to prolonged disease-free status, can be achieved in animals and in humans.[25-28,32] The mechanism by which protein A functions in these perfusion experiments has not been resolved, although the studies indicate that *S. aureus* was retained in the filter systems and therefore either the plasma was altered as it passed the bacteria or a bacterial product was released during perfusion. Hellström and Hellström[36] have suggested that, for example, protein A may leach off the *S. aureus*, enter the bloodstream, and induce changes in vivo leading to tumor responses. Soderberg et al.[37] have reported that large immune complexes form in the blood of rats immediately after intravenous injection of protein A. These large immune complexes are subsequently concentrated in the spleen. Immune complexes in the peripheral blood returned to pretreatment levels within 1-2 days.[37]

The repeated observation that immune complexes rose immediately after perfusion[25,26,29] suggested to us that protein A indeed leached from the filter systems during perfusion. We attempted to answer the question of protein A's ability to produce abrupt increases in circulating immune complexes and to induce tumor regressions through experiments involving direct intravenous injection of low concentrations of protein A.[38] Toxicity of protein A was also assessed. Dogs with histologically proven malignancy were obtained from breeders or private individuals. Measurable tumor was a study requirement. Most animals had undergone previous biopsy of tumor masses, though no dog was given chemotherapy or irradiation therapy prior to or during the trial. Following a 1 to 4-week period to verify progressive tumor growth, each animal was treated weekly with intravenous infusions of purified protein A. Therapy began at 0.06 mg/kg body weight of purified protein A and was continued at this dose for four weekly injections. The protein A dose was escalated by 0.06 mg/kg body weight every fourth week up to 36 weeks. All injections were administered in saline over 5–15 minutes and all animals were observed overnight for toxicity following each injection. Tumor measurements were obtained weekly just prior to each injection.

Thirteen animals were referred and considered acceptable for study.[38] Three were diagnosed with mammary adenocarcinoma, four with aggressive large cell lymphoma, two with malignant melanoma, and an additional four with various sarcomas. Four dogs demonstrated a decrease in tumor size, two had a greater than 50% decrease in the product of the two largest perpendicular diameters (partial response), and two additional dogs had less than a 50% decrease. Responses occurred at doses of 0.06 to 0.18 mg/kg body weight. All other animals displayed progressive disease. As the dose was increased, tumor growth progressed in all animals including the responders.

We have expanded on these studies by injecting a separate set of eight mammary carcinoma-bearing dogs with purified protein A at a single dose level of 0.06 mg/kg body weight (unpublished observations). Purified protein A was injected at intervals of either one, two, three, or five times per week in each of two dogs per injection schedule. Results of these injections were comparable to those observed with the escalating dose trial [38] in that tumor response representing up to 50% reduction in tumor size were observed in three of the eight dogs. One dog, infused five times per week, had extensive necrosis of its ulcerated tumor within the first week following initiation of treatment. These data indicate that while increased frequency of administration may intensify or shorten the time to an observed tumor response, the number of responding dogs remains the same as seen in the escalating dose trial.[38]

Toxicity from intravenous adminstration of purified protein A was not observed at the lowest dose (0.06 mg/kg). However, the frequency and severity to toxicity increased with increasing dose.[38] Between 0.12 mg/kg and 0.42 mg/kg, 30–50% of infusions were associated with toxicity. Lethargy was seen at low doses, and with dose escalation, nausea, vomiting, diarrhea, and micturation increased in severity. One animal had a seizure at 0.30 mg/kg. Panting was observed in 20–50% of the infusions less than 0.48 mg/kg, but above this level respiratory distress was noted in all procedures. Nausea, vomiting, and diarrhea were more severe than at lower doses.[38]

Hematologic abnormalities were noted above 0.48 mg/kg and also appeared to be dose-related. Immediately following protein A infusion, total white cell counts decreased. Lymphocytes fell to 1–3% of the circulating nucleated cells. Granulocytes and band forms began to increase by 1 hour and reached levels above baseline by 4–6 hours postinfusion of purified protein A. Lymphocytes did not return to baseline for 48–72 hours. Mild decreases in platelet counts were also observed immediately following the purified protein A infusion, but returned to baseline by 12–24 hours. No abnormalities in the red cell numbers or cumulative hematopoietic toxicity was recognized.[38]

Immune complex levels, as determined in Clq binding

assays, increased abruptly, immediately after intravenous infusion of protein A.[38] The sharply elevated immune complex levels persisted for 24–48 hours before returning to pretreatment levels. Circulating immune complex levels decreased over longer periods of time in concordance with tumor regression. During periods of tumor growth, circulating immune complex levels remained elevated.[38]

We conclude that protein A has anticancer properties when injected intravenously. However, the amount of protein A required to elicit a response was low (0.06 to 0.18 mg/kg body weight). In counterdistinction to removing immune complexes, these small quantities of protein A immediately increased the measurable circulating immune complexes. Our findings[38] support interpretations of others[27,29] that direct removal of circulating immune complexes by extracorporeal plasma perfusion may not be the primary mechanism of action leading to the demonstrated tumor responses. An alternative interpretation of these plasma perfusion studies is that small quantities of protein A may elute from *S. aureus* or from protein A bound matrices. The free protein A would then enter the bloodstream and directly alter tumor immune responses through an as yet undetermined mechanism. Terman et al.[26] have suggested that complement-mediated pathways are involved in the tumor regression process. Activation of alternate complement pathways by intravenous protein A injections is consistent with our findings, although data to support this is lacking.[38]

SUMMARY AND FUTURE DIRECTION

The mechanism of action of the observed responses and toxicity in animals and humans is not completely understood. The biologic effects of protein A have recently been reviewed by Langone.[13] Protein A added to serum can cause a marked depletion of complement activity. This may play a role in the observed toxicity as well as in tumor response. Also, histamine is released when white blood cells ingest protein A-IgG complexes. These complexes have also been implicated in serotonin release.[13] Over the last decade, it has become clear that protein A has mitogenic activity, activating both T and B cells.[13] In addition, protein A has been reported to stimulate the production of type II interferon.[39] It is possible that the tumor responses observed are mediated by generalized immune stimulation (T and B cells) and/or by the production of interferon, although no data is available to support this possibility.

Though the initial theory for the use of protein A was to remove immune-suppressive IgG complexes by binding at the Fc receptor protein A, it is now clear that other mechanisms of action can lead to tumor response. Responses have been observed with very small amounts of plasma treated with and without measurable circulating immune complexes. Decreases in tumor size have been observed in animals and in humans treated with various preparations of *S. aureus* that contained protein A on the surface or with preparations containing purified protein A adherent on other solid matrices. Since intravenous purified protein A has tumoricidal effects,[38] it is possible that protein A leaching off the ex vivo chambers causes the observed tumor responses. Responses following intravenous injection of protein A were observed at low doses, and no evidence exists from our studies that a direct protein A dose-response relationship exists. In fact, tumor growth occurred in all animals at higher doses. Responses occurred at 0.06 mg/kg, 0.12 mg/kg, and 0.18 mg/kg in mammary adenocarcinoma, melanoma, lymphoma, and osteosarcoma. Toxicity was not observed or was minimal at these low doses and increased with an increase in protein A dose. Complement activation may explain this toxicity.[38]

From our studies, we hypothesize that large immune complexes form in the blood following protein A injections and that these complexes activate complement through an alternate pathway.[13] This activated complement may combine with tumor-associated antibodies on the surface of tumor cells leading to complement-dependent antibody cytotoxicity. It is also possible that this complement activation can be initiated in the extracorporeal perfusion chambers and that tumor responses are generated by this activated complement following return of the plasma to the tumor-bearing individual.

Future studies are being directed toward identifying the mechanism by which protein A acts on tumor immunity and toward determining the optimal treatment regimens that will elicit the maximum responses. Concurrent with these studies are needed investigations to identify types of human tumors where this therapy will be beneficial.

REFERENCES

1. Verwey WF: A type-specific antigenic protein derived from the staphylococcus. *J Exp Med* 1940; 71:635.
2. Jensen K: A normally occurring staphylococcus antibody in human serum. *Acta Pathol Microbiol Scand* 1958; 44:421.
3. Jensen K: The toxic effect of a staphylococcal polysaccharide on isolated guinea pig ileum and the antitoxic effect of normal human serum. *Acta Allergol* 1959; 13:89.
4. Grov A, Myklestad B, Oeding P: Immunochemical studies on antigen preparations from *Staphylococcus aureus*. 1. Isolation and chemical characterization of antigen A. *Acta Pathol Microbiol Scand* 1964; 61:588.
5. Oeding P, Grov A, Myklestad B: Immunochemical studies on antigen preparations from *Staphylococcus aureus*. 2. Precipitating and erythrocyte-sensitizing properties of protein A (antigen A) and related substances. *Acta Pathol Microbiol Scand* 1964; 62:117.
6. Forsgren A, Sjöquist J: "Protein A" from *S. aureus*. I. Pseudoimmune reaction with human γ-globulin. *J Immunol* 1966; 97:882.
7. Kronvall G, Seal US, Svensson S, et al: Phylogenetic aspects of staphylococcal protein A-reactive serum globulins in birds and mammals. *Acta Pathol Microbiol Immunol Scand* [B] 1974; 82:12.
8. Kronvall G, Williams RC Jr: Differences in antiprotein A activity among IgG subgroups. *J Immunol* 1969; 103:828.
9. Harboe M, Fölling I: Recognition of two distinct groups of human IgM and IgA based on different binding to staphylococci. *Scand J Immunol* 1974; 3:471.
10. McDowell G, Grov A, Oeding P: IgA and IgM reacting with staphylococcal protein A. *Acta Pathol Microbiol Immunol Scand* [B] 1971; 79:801.
11. Kronvall G, Seal US, Finstad J, et al: Phylogenetic insight into evolution of mammalian Fc fragment of γG globulin using staphylococcal protein A. *J Immunol* 1970; 104:140.

12. Langone JJ, Boyle MDP, Borsos T: A solid-phase immunoassay for human immunoglobulin E: Use of ^{125}I-labeled protein A as the tracer. *Ann Biochem* 1979; 93:207.

13. Langone JJ: Protein A of *Staphylococcus aureus* and related immunoglobulin receptors produced by streptococci and pneumonococci. *Adv Immunol* 1982; 32:157.

14. Bansal SC, Bansal BR, Boland JP: Blocking and unblocking serum factors in neoplasia. *Curr Top Microbiol Immunol* 1976; 75:45.

15. Hellström I, Hellström KE: Studies on cellular immunity and its serum-mediated inhibition in moloney-virus-induced mouse sarcomas. *Int J Cancer* 1969; 4:587.

16. Bansal SC, Hargreaves R, Sjögren HO: Facilitation of polyoma tumor growth in rats by blocking sera and tumor eluate. *Int J Cancer* 1972; 9:97.

17. Jose DG, Seshadri R: Circulating immune complexes in human neuroblastoma: Direct assay and role in blocking specific cellular immunity. *Int J Cancer* 1974; 13:824.

18. Klein G: Immune and nonimmune control of neoplastic development: Contrasting effects of host and tumor evolution. *Cancer* 1980; 45:2486.

19. Sjögren HO, Hellström I, Bansal SC, et al: Suggestive evidence that the "blocking antibodies" of tumor-bearing individuals may be antigen-antibody complexes. *Proc Natl Acad Sci USA* 1971; 68:1372.

20. Baldwin RW, Price MR, Robins RA: Blocking of lymphocyte-mediated cytotoxicity for rat hepatoma cells by tumour-specific antigen-antibody complexes. *Nature New Biol* 1972; 238:185.

21. Baldwin RW, Price MR, Robins RA: Significance of serum factors modifying cellular immune responses to growing tumours. *Br J Cancer* 1973; 28(Suppl) 1:37.

22. Isbister WH, Noonan FP, Halliday WJ, et al: Human thoracic duct cannulation: Manipulation of tumor-specific blocking factors in a patient with malignant melanoma. *Cancer* 1975; 35:1465.

23. Israel L, Edelstein R, Mannoni P, et al: Plasmapheresis in patients with disseminated cancer: Clincial results and correlation with changes in serum protein. The concept of "nonspecific blocking factors." *Cancer* 1977; 40:3146.

24. Steele G Jr, Ankerst J, Sjögren HO: Alteration in in vitro anti-tumor activity of tumor-bearer sera by absorption with *Staphylococcus aureus*, Cowan I *Int J Cancer* 1974; 14:83.

25. Bansal SC, Bansal BR, Thomas HL, et al: Ex vivo removal of serum IgG in a patient with colon carcinoma: Some biochemical, immunological, and histological observations. *Cancer* 1978; 42:1.

26. Terman DS, Yamamoto T, Mattioli M, et al: Extensive necrosis of spontaneous canine mammary adenocarcinoma after extracorporeal perfusion over *Staphylococcus aureus* Cowans 1. I. Description of acute tumoricidal response: Morphologic, histologic, immunohistochemical, immunologic, and serologic findings. *J Immunol* 1980; 124:795.

27. Messerschmidt G, Bowles C, Alsaker R, et al: Long-term follow-up of dogs with spontaneous mammary tumors treated with ex vivo plasma

perfusion over *S. aureus* Cowan strain I (SAC). *Proc Am Assoc Cancer Res* 1982; 23:279 (abstract).

28. Holohan TV, Phillips TM, Bowles C, et al: Regression of canine mammary carcinoma after immunoadsorption therapy. *Cancer Res* 1982; 42:3663.

29. Jones FR, Yoshida LH, Ladiges WC, et al: Treatment of feline leukemia and reversal of FeLV by ex vivo removal of IgG: A preliminary report. *Cancer* 1980; 46:675.

30. Terman DS, Yamamoto T, Tillquist RL, et al: Tumoricidal response induced by cytosine arabinoside after plasma perfusion over protein A. *Science* 1980; 209:1257.

31. Messerschmidt G, Bowles C, Dean D, et al: Phase I trial of *Staphylococcus aureus* Cowan I immunoperfusion. *Cancer Treat Rep* 1982; 66:2027.

32. Terman DS, Young JB, Shearer WT, et al: Preliminary observations of the effects on breast adenocarcinoma of plasma perfused over immobilized protein A. *N Engl J Med* 1981; 305:1195.

33. Messerschmidt G, Steis R, Dowling R, et al: Phase I trials of ex vivo plasma perfusion with *S. aureus* Cowan strain I (SAC) and protein A charcoal (PACC) in malignancy. *Proc Am Soc Clin Oncol* 1982; 1:80 (abstract).

34. Messerschmidt GL, Bowles C, Parrillo J, et al: The use of protein A in the treatment of malignancy: Rationale and the NCI experience. *Prog Clin Biol Res* 1982; 106:385.

35. Bensinger WI, Kinet JP, Hennen G, et al: Plasma perfused over immobilized protein A for breast cancer. *N Engl J Med* 1982; 306:935 (letter to the editor).

36. Hellström KE, Hellström I: Does perfusion with treated plasma cure cancer? *N Engl J Med* 1981; 305:1215 (editorial).

37. Soderberg FB, Siag WM, Jones JM: Alteration by protein A of in vivo distribution of immune complexes containing antigens of MuLV retrovirus. *Fed Proc* 1982; 41:324 (abstract).

38. Messerschmidt G, Bowles C, Alsaker R, et al: Toxic, immunologic, and pathologic changes in dogs with spontaneous neoplasia treated with IV infusions of purified staph protein A (SPA). *Fed Proc* 1982; 41:325 (abstract).

39. Catalona WJ, Ratliff TL, McCool RE: γ interferon induced by *S. aureus* protein A augments natural killing and ADCC. *Nature* 1981; 291:77.

CHAPTER 6

Immunoadsorption with Staphylococcal Protein A in Human Neoplasms

Jean-Pierre Kinet, M.D.
William I. Bensinger, M.D.
Francis Frankenne, Ph.D.
Georges Hennen, M.D., Ph.D.
Jacqueline Foidart, M.D., Ph.D.
Philippe Mahieu, M.D., Ph.D.

INTRODUCTION

Staphylococcus aureus, Cowan strain I, has been shown to contain a cell wall substance, protein A, which specially binds to the Fc fragment of certain subclasses of IgG and, to a lesser extent, IgM, from most mammals.[1] This selective reactivity has been the basis of various immunoassays, in which ^{125}I-labeled protein A of high specific and functional activity serves as a general tracer for the detection of antibodies and, indirectly, of antigens.[2,3]

Different cellular Fc receptors react more avidly with complexed IgG than with free immunoglobulin G because of the potential for multipoint attachment.[4,5] It has been shown[6] that the Fc receptor of protein A also binds complexed IgG with a greater affinity than monomeric IgG. A *Staphylococcus aureus* binding assay for aggregated IgG and immune complexes in human sera has been devised and demonstrated to be specific, reliable, quantitative, not complement-dependent, and as sensitive as two complement-dependent assays, i.e., the Clq binding assay and the Raji cell assay.[6]

Since protein A has the biological property of interacting with and binding to the Fc portion of IgG, and since it preferentially reacts with antigen-antibody complexes in vitro, one may therefore wonder whether this molecule might also be useful for the ex vivo removal of IgG antibody and/or of immune complexes from the serum of patients presenting with either an autoimmune disease, an immune complex-mediated disease, or a malignant disease in which circulating immune complexes ("blocking factors") are detectable. Since we mainly have used protein A columns in human neoplasms, we will describe in this chapter our experience in this field. Five major sections will be successively described:

    (1)   the rationale and background for using ex vivo perfusion of plasma over protein A in patients with cancer;

    (2)   the preparation of the immunosorbent, the description of the extracorporeal perfusion system used and the patients treated;

    (3)   the safety and efficacy of the procedure;

    (4)   the possible mechanisms involved in the side effects and tumoricidal response;

    (5)   the conclusions and future developments.

RATIONALE AND BACKGROUND TO USE EX VIVO PLASMA PERFUSION OVER PROTEIN A IN MALIGNANT DISEASES

Previous studies have shown that most malignancies possess tumor-associated antigens capable of evoking both humoral and cell mediated immune response in their hosts. Precipitating antibodies have been demonstrated in a variety of human tumors including colon carcinoma,[7] osteosarcoma,[8] melanoma,[9] Burkitt's lymphoma,[10] neuroblastoma,[11] bladder tumor,[12] and Hodgkin's disease.[13] Cell-mediated immunity has also been demonstrated for neuroblastoma,[14] colon carcinoma,[15,16] bladder carcinoma,[12,17,18] breast adenocarcinoma,[19,20] lung and kidney carcinoma.[20]

The significance of this humoral and cell-mediated immunity remains unknown since by the time malignancies are clinically detectable, most grow progressively, killing the host in spite of any demonstrated immune reaction directed against the growing malignant cells. The discovery of a "blocking factor"[21] suggested one possible explanation to this paradox. Rabbits carrying Shope virus-induced papillomas have been shown to have lymphocytes capable of killing papilloma tumor cells in culture in the absence of added autologous serum. Their serum was shown to contain a substance termed "blocking factor," which is able to inhibit the lymphocyte cytotoxic response. Sera from rabbits in which the tumor regressed did not block this response.[21] In mice, it has been shown that the tumor growth was enhanced when animals were injected with serum containing blocking factor after tumor isografting, and that the tumor growth was strongly inhibited when mice were injected with immune serum lacking blocking factors after tumor isografting.[22,23] Steele et al.[24] have demonstrated that protein A could remove the blocking activity from the sera of tumor-bearing animals. The nature of blocking factor remains controversial. It has been variously thought to be antibody,[22,25,26] antigen,[27–29] or antigen-antibody complexes.[30,31] Indeed, at least one investigator has found evidence for each of these possible mechanisms within one single model, the rat hepatoma.[32]

There is some evidence to indicate that blocking factor might also play a role in malignant disease in man and that modification of the patient's plasma might be of clinical benefit. Hellström et al.[33] have sequentially followed 10 patients with malignant melanoma by measuring lymphocyte cytotoxicity to both autologous and cultured melanoma cell lines, and the ability of the patients' autologous sera to block this response. The results demonstrated that patients with little or no residual tumor had stronger lymphocyte cytotoxicity and decreased levels of serum blocking factor than patients with high tumor loads.[33] Similar results have been obtained in human neuroblastoma.[31]

An important question that remains is how these and similar studies by other investigators[34,35] can be translated into a system that will allow a successful immunological manipulation of a host so that a beneficial tumoricidal effect might be obtained. Several nonspecific clinical approaches to dealing with "blocking factors" have been reported in preliminary work.[36–39] Thoracic duct drainage in a patient with advanced melanoma resulted in both a decrease in serum "blocking factor" and a temporary clinical remission. In this case, the cell-free lymph only was removed, the washed lymphocytes being returned to the patient.[38] Israel et al.[39] performed plasma exchanges in 23 patients with disseminated cancer. Eight of 23 had partial responses without any other changes in therapy. These patients had a variety of tumors, including breast, colon, melanoma, and thyroid carcinoma. Hersey et al.[37] performed plasma exchanges in four patients with melanoma. Three of four patients demonstrated an increase in antibody-dependent cell-mediated cytotoxicity. Other investigators have reported similar results with plasma exchanges.[36]

Different methods for the specific ex vivo removal of serum mammalian IgG have been subsequently described.[40–44] Bansal et al.[40] have used heat-killed, formalin-fixed *Staphylococcus aureus* in an ex vivo plasma perfusion system with on-line micropore filtration-sterilization of perfused plasma. They first demonstrated the feasibility of this technique in normal dogs.[40] They then reported its use in a single patient with advanced colon carcinoma as the only form of therapy.[41] Healing of ulcerated tumor was noted after three treatments. The patient tolerated a total of 20 such treatments over a 6-month period, and a follow-up laparotomy showed minimal evidence of active tumor. Multiple biopsies of this tumor demonstrated major fibrosis.[41] Terman et al.[44] reported on whole *Staphylococcus* plasma perfusion in 12 dogs with advanced mammary carcinoma. Dogs "perfused" with *S. aureus* Cowan I had healing of tumors, whereas dogs "perfused" with *S. aureus* Wood 46, which does not contain protein A, had no improvement. This experiment suggests that protein A is one of the necessary components for antitumor effect. Other investigators have reported similar results using whole staphylococcus or protein A affinity columns.[43,45] Cats spontaneously infected with feline leukemia virus and with evidence of active leukemia have been treated with a combination of low dose irradiation and extracorporeal immunoadsorption, using formalin-fixed and heat-killed *S. aureus*.[42] The treatment resulted in a reduction of circulating lymphoblasts within 2 weeks and a clinical improvement in three of the five animals, in contrast to the two cats receiving irradiation therapy only who died of their disease within 6 weeks.[42]

More recently, Ray et al.[46] have used whole *S. aureus* plasma perfusion to successfully treat a multiple myeloma patient with hyperviscosity syndrome due to abnormal IgG, and Terman et al.[47] observed objective partial remission or improvement in four patients presenting with a breast adenocarcinoma after perfusion of their plasma over purified protein A immobilized in a collodion-charcoal matrix.

These clinical and experimental results therefore support the concept described some 12 years ago by Hellström et al.[48] and demonstrate that circulating "blocking factors" could suppress in animals and patients with cancer the ability of their own lymphocytes to destroy tumor cells in vitro. They further suggest that some of these "blocking factors" could be removed from the serum by ex vivo perfusion over protein A columns and therefore provide a theoretical basis for using this procedure in patients with disseminated cancer.

However, it should be noted that protein A is known to have other important immunologic effects in vitro including activation of T and B cells,[49] induction of interferon production,[50] and activation of complement.[51] Any or all of these effects may play a role in the tumoricidal effect of protein A.

PURIFICATION OF PROTEIN A: DESCRIPTION OF THE EXTRACORPOREAL PERFUSION SYSTEM

Staphylococcus aureus, Cowans strain I, is grown as a batch culture in a stirred fermentor with a working volume of 10 liters. The culture medium and the conditions for growth are the same as those previously described.[52] The bacteria (300 g) are treated by lysostaphin (15 mg), and after 2 hours protein A is classically purified from the supernatant by affinity chromatography over a human IgG-Sepharose column.[52] The recovered protein is cell-free and contains greater than 99% protein A, as shown by analytical methods[52] and by a competitive inhibition solid phase radioimmunoassay (RIA).[2,53] The principle of this RIA is the same as that described by Langone et al.,[2] except that human IgG is bound to polyvinylchloride plates (Falcon 3912 Microtest III plates, Becton Dickinson, Oxnard, CA) instead of agarose beads (Bio-Rad Laboratories, Rockville Centre, NY). The amount of IgG coated was approximately 5 μg per microtiter well.[53] Protein A purchased from Pharmacia (Uppsala, Sweden) was used as an internal standard to control the purity of our protein A.[53]

In order to determine whether the Fc receptor binding activity of protein A is involved in the potential tumoricidal response, the Fc receptor binding capacity of protein A was destroyed either by heating protein A for 0.5 hour at 100°C (1 mg/ml in phosphate-buffered saline, pH 7.4) or by treatment with 20% polyethylene glycol (or 20% dimethyl sulfoxide).[53,54] It has been shown that the four tyrosine residues of protein A seem to be of crucial importance for the reactivity to the Fc region of immunoglobulins.[54] Of the four tyrosine residues, 3.5 are perturbed by 20% polyethylene glycol, whereas all of them are perturbed by 20% dimethyl sulfoxide[54] or by heating at 100°C for 30 minutes.[53] After this latter treatment, the binding of protein A to IgG-coated microtiter plates is completely removed, as monitored by competitive inhibition solid phase RIA using ^{125}I-labeled protein A as a tracer.[53] The material, however, partly retains its capacity to be bound, either by goat antibodies to protein A or by their F (ab')$_2$ fragments.[53]

Both native and denatured protein A are covalently linked by carbodiimide to crystalline silica according to the method described by Weetall in 1976.[55] This linkage is stable and not easily degraded on exposure to plasma. Furthermore, since this chemical reaction uses the lysine residues of protein A, it will not interfere with the interaction of native protein A with the Fc domain of IgG. Each column contains 200 mg of protein A linked to 100 g of silica. The protein A columns have been used in conjunction with a continuous flow membrane separator (Plasmaflo Asahi, Tokyo, Japan or Travenol Company, Lessines, Belgium), with a cut-off of about 3×10^6 daltons. These membranes have been used as previously described[56] after placement either of catheters into the subclavian or femoral veins, or of a temporary arteriovenous shunt. Separated plasma is then perfused through the protein A column at a flow rate of 20 to 30 ml per minute and returned to the patient with the formed blood elements (Fig. 1). All patients are treated eight to 13 times once or twice a week, and during each session 3 liters of plasma are passed through the protein A column. During the procedure, anticoagulation is performed by intravenously injecting a pulse of 5,000 IU of heparin, followed by a continuous infusion of 2,000 IU heparin per hour. At the end of the procedure the column is washed with 2 liters of 0.01 M phosphate-buffered saline, pH 7.4. The material bound to the protein A is eluted with 0.5 liters of 3M KSCN, and the immunoadsorbent is extensively washed with 3 liters of sterile saline and then stored at 4°C until the next use in a 0.02% sodium azide solution in sterile saline.

The immunoadsorbent support is a crystalline form of diatomaceous earth commonly utilized in gas chromatography systems. It has been used in humans for removal of anti-ABO antibodies with no demonstrable toxicity.[57]

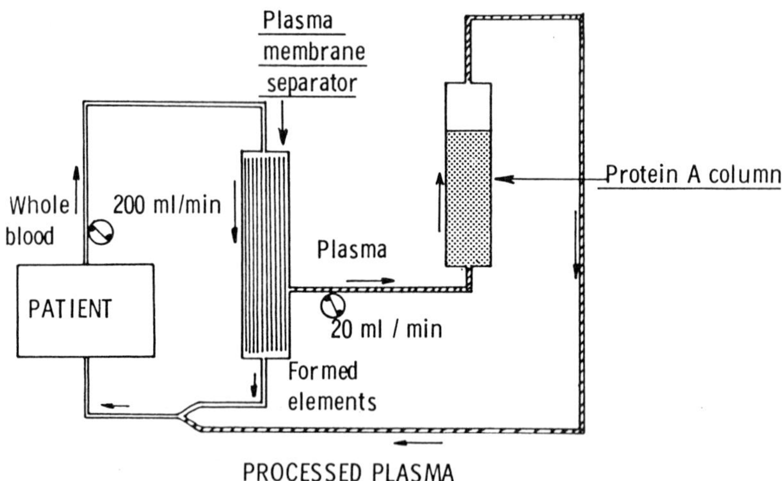

Figure 1: *Diagrammatic representation of the extracorporeal perfusion system used for perfusion of plasma over protein A columns in patients with disseminated cancer.*

PATIENTS

Since one of our major objectives was to ascertain a tumoricidal response, the patients were treated by protein A immunoadsorption only during the experimental protocol and were generally selected on the basis of the presence of easily measurable primary tumor masses and/or metastatic lesions. Each patient received, as objectively as possible, information on the protocol followed, knew its experimental nature, and gave informed oral consent according to the criteria of the ethical committee of the University of Liège. The main clinical features of the patients are summarized in Table 1.

In all of the patients studied, the diagnosis was based on the histological analysis of biopsy specimens. The primary tumors were adenocarcinomas of the breast (nine cases), colon (one case), and pancreas (one case). Two patients never received any therapy before the protein A immunoadsorption

Table 1
Initial Status of the Patients

Patient	Sex	Age (yrs)	Primary tumor	Extension	Previous therapy
Group A[a]					
1	F	44	Breast	Bone, liver, lymph nodes	None
2	F	50	Breast	Lymph nodes, bone	Mastectomy, radiotherapy, hormonotherapy, chemotherapy[c]
3	F	54	Breast	Lymph nodes	None
4	F	59	Breast	Lung, bone	Mastectomy, hormono-therapy, chemotherapy[d]
5	F	43	Breast	Bone	Mastectomy, hormono-therapy, radiotherapy
6	M	61	Pancreas	Skin	Surgery
7	F	32	Breast	Skin, bone	Mastectomy, radiotherapy
8	F	42	Breast	Bone, lung	Mastectomy, radiotherapy, hormonotherapy
Group B[b]					
9	F	55	Breast	Bone	Mastectomy, radiotherapy
10	F	58	Breast	Bone, liver	Mastectomy, hormono-therapy, radiotherapy
11	M	51	Colon	Liver, lymph nodes	Surgery, chemotherapy[e]

[a]Plasma perfused over columns of "native" protein A.
[b]Plasma perfused over columns of "denatured" protein A.
[c]Cyclophosphamide, methotrexate, 5-fluorouracil, vincristine, adriamycin.
[d]5-Fluorouracil, adriamycin, vincristine.
[e]5-Fluorouracil.

whereas the nine others were refractory to conventional treatments including surgery, radiation therapy, hormonotherapy, and/or chemotherapy before the immunoadsorption. In these cases, a delay of at least 1 month elapsed between the last therapy and the first immunoadsorption procedure. In order to study toxic effects of the protein A columns, each patient had a complete physical examination, a complete blood count, a study of liver and renal function, a chest roentgenogram, an electrocardiogram, and in some cases echocardiography before starting the protein A sessions. During the treatment, frequent monitoring of vital signs, including blood pressure, pulse, temperature, electrocardiogram and blood cultures were performed.

In order to ascertain the tumoricidal effect of protein A immunoadsorption, sequential measurements of tumor sizes were followed by these techniques: mammography, thermography, metastatic skeletal survey, radionuclide scans of the liver, spleen, and bones, abdominal echotomography, and chest roentgenogram. Tumor biopsies were also performed before and after protein A immunoadsorption in five patients (cases 2, 3, 6, 9, and 11).

In order to determine whether the Fc receptor-binding activity of protein A was involved in the eventual tumoricidal response, the plasma of patients from group A (cases 1 to 8) was perfused over columns of "native" protein A, i.e., protein A exhibiting a normal Fc receptor-binding activity, and the plasma of patients from group B (cases 9 to 11) perfused over columns of "denatured" protein A, i.e., protein A in which the Fc receptor-binding activity was destroyed.[53,54]

Acute Side Effects of Plasma Perfusion over Protein A Columns

None of the patients from group B ever developed any major side effects, in contrast to most patients from group A. Indeed, in these latter patients, episodes of hypotension and tachycardia were frequently observed and easily corrected by infusion of isotonic saline (100 to 200 ml of saline over 30 minutes). More severe toxicity was noted in patients 1, 3, 5, 6, and 7 who exhibited chills corrected by intravenous injection of 50 mg of meperidine. Patient 5 became flushed and developed urticaria during the first session. Administration of Benadryl (100 mg IV in two divided doses) markedly reduced the severity of these symptoms.

Patient 8, who had extensive lung metastases, developed after the second session a severe bronchospasm and an acute respiratory distress syndrome. She had dyspnea, hypoxemia, reduced respiratory system compliance, and diffuse alveolar infiltrates that resembled pulmonary edema. She died 2 weeks later from refractory hypoxemia. Since severe bronchospastic reactions were never observed in patients without lung involvement, patients with pulmonary metastases may not be suitable candidates for this procedure. Similar toxic pulmonary reactions have also been reported by Terman et al. in one patient after perfusion of plasma over protein A immobilized in a collodion-charcoal matrix. This patient also had lung metastases.[47]

Most patients of group A exhibited moderate to severe pain localized in tumor sites. This pain generally started during the session, lasted for 1 to 2 days, and was frequently associated with temperature elevation. Systematic investigation for bacterial infection in blood and urine samples was always negative. Liver and renal function also remained unchanged during the immunoadsorption procedure.[58,59]

EFFECTS OF PLASMA PERFUSION OVER PROTEIN A ON SOME CLASSICAL BLOOD PARAMETERS

In patients from groups A and B, electrolyte, red blood cell, and platelet counts remained unchanged during and after the immunoadsorption over protein A. In patients from group A, leukocyte counts decreased by about 40% 30 minutes after the beginning of the procedure. One hour later leukocyte counts increased to reach maximum values ranging between 20,000 to 30,000 cells/mm^3 (Fig. 2) (containing more than 90% polymorphonuclear leukocytes). The leukocyte counts returned to the pretreatment levels 2 to 3 days later.[10,26] In patients from group B, leukocyte counts decreased by about 30% 30 minutes after the beginning of the procedure but returned to the pretreatment levels at the end of the procedure.

In patients from group A, IgG and C3 plasma levels decreased by about 30% in the postperfusion period, return-

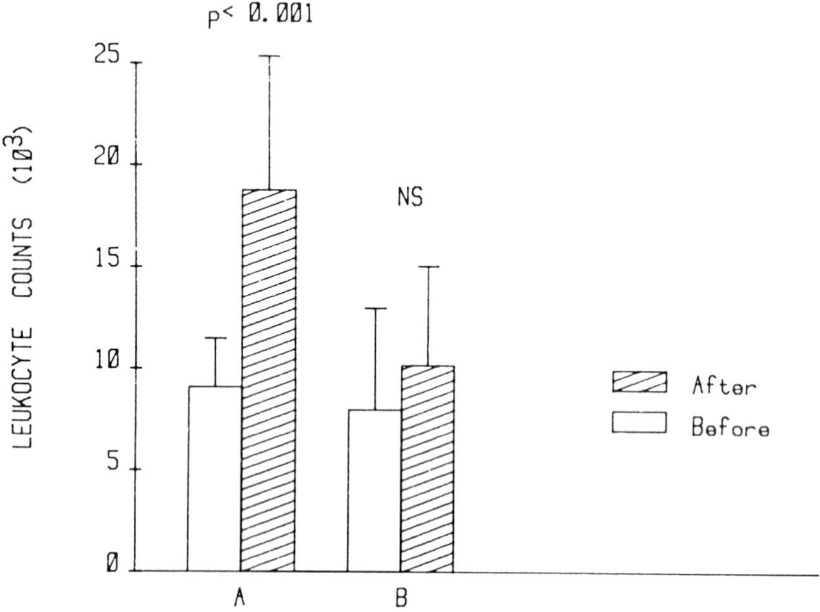

Figure 2: *Leukocyte counts before and after ex vivo perfusion of plasma over protein A columns in patients with disseminated cancer. A = columns of "active" protein A; B = columns of "denatured" protein A.*

ing to (or exceeding) the pretreatment levels 48 to 72 hours later.[10,26] IgA and IgM plasma levels also decreased, but to a lesser degree than IgG and C3. The decrease in IgA and IgM plasma levels was probably related to some blood dilution by the saline infused to correct the hypotensive episodes. In patients from group B, the IgG, IgM, IgA, and C3 plasma levels all were reduced by the same degree, i.e., about 10%. It is believed that, in these patients, the fall in IgG, IgM, IgA, and C3 plasma levels was due either to some serum dilution caused by priming solution in the apparatus or to a nonspecific adsorption of the components on the membrane filters.[51,58,59]

Taken together, these results demonstrate that most biochemical, hematological, and immunological parameters measured after plasma perfusion over protein A are not strongly altered, and that, when present, these alterations are generally reversible and mainly related to some interactions between plasma components and the Fc receptors of protein A.

EFFECTS OF PLASMA PERFUSION OVER PROTEIN A ON THE CLINICAL COURSE OF THE DISEASE

Table 2 summarizes the subjective and objective effects observed in both groups of patients. In group A, four patients exhibited an objective partial regression after two to five sessions and before any other therapy had been undertaken. Two patients had stable disease, one exhibited an increased size of one bone metastatic lesion and one developed severe refractory hypoxemia. Except for patient 8, there was an improvement in the general condition of the patients after two to four sessions, as evidenced by a decrease of pain in tumor sites and a subjective feeling of well-being. Some patients spontaneously reduced their analgesic drugs. In group B, all the patients exhibited progression of their disease during the procedure.

The patients were followed for 1 to 15 months after the first treatment (Table 2). In patient 1, the product of perpendicular diameters in the primary tumor and in one lymph node showed a 50% decrease after five sessions, thus indicating a partial remission according to the criteria of the "International Union Against Cancer."[60] After eight perfusions, she received cytotoxic drugs, i.e., adriamycin, vincristine, cyclophosphamide, methotrexate, and 5-fluorouracil. She is currently in remission on the basis of the following criteria: (1) the primary breast tumor is now undetectable by physical examination, mammography, and thermography; (2) the local adenopathies have completely disappeared; (3) the metastatic bone lesions are still present on the roentgenograms but the hyperactive bone areas shown in the first radionuclide scans are now much less obvious.

Patient 2 had a great number of metastatic lesions in

Table 2
Effects of Protein A Immunoadsorption

Patient	Perfusion No.	Weight (kg) Before	Weight (kg) After	Clinical effects Subjective	Clinical effects Objective	Histological changes of tumors	Follow-up (months)
Group A[a]							
1	8	44	46	Disappearance of bone pain	Partial regression	—	15
2	8	50.7	53.6	None	Partial regression	Necrosis-fibrosis	1½[c]
3	13	71	74	Felt well	Partial regression	Necrosis-fibrosis	10
4	8	44	44	Decrease of dyspnea	No change	—	8
5	13	66.6	68.2	Increased intensity of bone pain	Progression of one lesion	—	6
6	12	49	48	Decreased consistency of cutaneous lesions	Partial regression	Necrosis	4
7	10	55	58	Felt well	No change	—	4
8	2	65	ND[d]	Increase of dyspnea	Refractory hypoxemia	—	1
Group B[b]							
9	13	56.5	56.5	None	Progression	Numerous invading cells	6
10	4	70	68	None	Progression	—	1
11	12	75	72	None	Progression	Numerous invading cells	5

[a]See Table 1.
[b]See Table 1.
[c]Died.
[d]ND = not determined.

bones, liver, and paraaortic abdominal lymph nodes. She was cachectic at the beginning of plasma perfusions and died 6 weeks later after eight treatments despite a 35% objective regression of the lymph node abdominal mass.

In patient 3, the primary breast tumor decreased by 25%, thus indicating a less than partial remission.[60] After the last treatment the breast was excised. The tumor appeared as a 5 × 6 cm fibrous nodule. She subsequently received the same chemotherapy as patient 1, and no evidence of recurrence has been noticed 10 months later. In patient 4, the size of nodular lung metastatic lesions remained unchanged; she is alive 8 months later without having received any other therapy.

In patient 5, a bone radionuclide scan performed after the series of treatments did not show any new hyperactive areas (in contrast to what occurred within the preceding 3-month period during which the patient was under hormonotherapy). A metastatic lesion located in the left femoral diaphysis showed an increased size but another located on the right costal rib showed recalcification by roentgenogram. Nevertheless, this patient was considered as a nonresponder to protein A immunoadsorption.

Patient 6 had a great number of cutaneous metastatic lesions. After five procedures, their number and size decreased. This patient is actually alive without any other therapy. Patient 7 exhibited one bone and numerous cutaneous metastatic lesions. After 10 treatments, no changes were noted. Patient 8 presented with metastatic lung lesions. After the second session, this patient developed an acute respiratory distress syndrome and died 2 weeks later from refractory hypoxemia.

In the three patients from group B, an evident progression of bone (case 9) or liver (cases 10 and 11) metastatic lesions occurred during the procedure. Although patients 10 and 11 died after 1 and 5 months, respectively, patient 9 is still alive after 6 months and is currently undergoing the same chemotherapy as patients 1 and 3.

EFFECTS OF PLASMA PERFUSION OVER PROTEIN A ON THE HISTOLOGICAL ASPECT OF TUMOR CELLS

In patients 2, 3, 6, 9, and 11, tumor biopsy specimens were taken after 2, 5, 8, 10, and 8 procedures, respectively. The primary breast (cases 2, 3, 9), colon (case 11), and pancreas (case 6) tumors were all adenocarcinomas. In patients from group A (cases 2, 3, and 6), consistent histological changes were noted, i.e., tumor cells in all stages of destruction were seen, sometimes associated with inflammatory cell infiltration and fibrosis. It is of interest that this necrolytic aspect was observed especially in the periphery of the tumor. However, focal areas of active tumor proliferation were always present.[58,59] In patients from group B, the tumor proliferation remained obvious.

In summary, these results clearly show that plasma perfusion over protein A can induce a tumoricidal response in some patients presenting with an adenocarcinoma and metastatic lesions. They are therefore in agreement with observations made by others,[47] using a different extracorporeal perfusion system. Furthermore, our results strongly suggest that the Fc receptor function of protein A plays a major role in the tumoricidal effect observed in vivo with human neoplasms. However, in our experience, we have never observed complete remission but only partial tumor regression after plasma perfusion over protein A.

POSSIBLE MECHANISMS INVOLVED IN THE ACUTE SIDE EFFECTS AND NECROLYTIC RESPONSE

Acute Side Effects

The plasma of the patients is first separated from the whole blood by a macromolecular filter before being perfused over protein A columns and subsequently reinfused into the bloodstream with the formed elements. The safety of macromolecular filters is now well recognized by many European centers that have used these filters for several years and have not reported major side effects.[56] Furthermore, it must be recalled that the acute side effects have been observed mainly in patients from group A, i.e., in patients treated by columns containing protein A with a "normal" Fc receptor function. Some acute side effects might be therefore related to an interaction between plasma components and the Fc receptor of protein A, these interactions taking place either inside the column itself or in the bloodstream, if some leaking of protein A occurs.

Evidence for Protein A Leaking

The covalent linkage between protein A and crystalline silica is not easily degraded on exposure to plasma.[53] A leaking of protein A during the procedure is therefore highly improbable. However, in order to ascertain this fact, we have developed an RIA for measurement of protein A in biological fluids and particularly in serum.[53] The principle of this solid phase RIA is illustrated in Figure 3A. Antibodies to protein A were coated to polyvinylchloride plates (Falcon 3912, Becton Dickinson, Oxnard, CA) as glomerular basement membrane antigens.[61] In order to establish the standard curve, increasing amounts (from 1 ng to 1μg) of protein A were mixed with 1 ml of undiluted and decomplemented normal human serum for 2 hours at 37°C. Fifty-microliter samples were then incubated with the anti-protein A antibody-coated plates for 4 hours at 37°C. After extensive washing with phosphate-buffered saline (PBS) (containing 0.1% Tween 20), 50 μl of ^{125}I-F(ab')$_2$ fragments of anti-protein A antibodies (labeled by the method of Bolton and Hunter)[62] were added to each microtiter well. After an overnight incubation at 4°C and extensive washing with the PBS-Tween solution, the samples were counted in a gamma counter.[61] Each sample was tested in triplicate. An example of the standard curve is shown in

Figure 3B. This assay is sensitive since it allows the detection of 10 ng of protein A per ml of undiluted serum. Furthermore, the results are highly reproducible and are not influenced either by the presence of the Fc domain of IgG molecules or by the Fc receptor of protein A.[53]

Using this RIA, we have searched for the presence of protein A in the serum of patients from groups A and B. The following observations were made:

(1) a leaking of protein A occurred in both groups of patients;

(2) the protein A serum concentrations ranged from 10 ng/ml to 150 ng/ml;

(A)

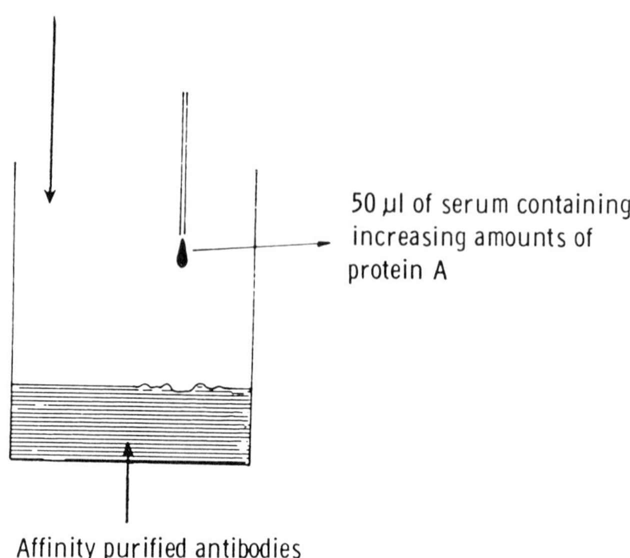

Figure 3: (A) Continued

(3)   major side effects, i.e., episodes of severe hypotension and tachycardia, chills, diarrhea, and temperature elevation of pulmonary rales, were more frequently observed in patients in whom a leaking of protein A was demonstrable by RIA than in patients in whom a protein A leaking was not demonstrable by this technique (Table 3);

(4)   despite evidence for protein A leaking, no major side effects were observed when the plasma of the

(B)

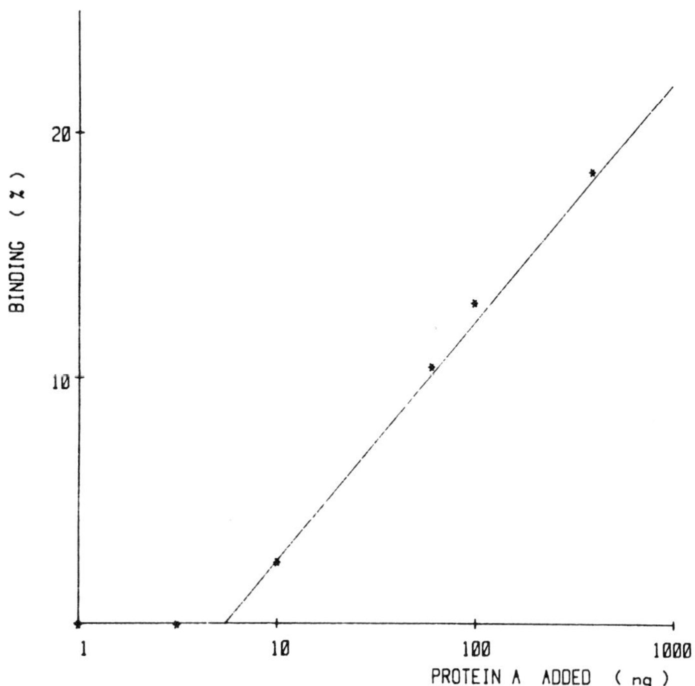

Figure 3: *(A)* *Schematic representation of the solid phase radioimmunoassay for measurement of protein A in normal human serum.*
(B) *Binding of* ^{125}I-$F(ab')_2$ *fragments of anti-protein A antibodies to unlabeled protein A as a function of the amount of protein A added.*

Table 3
Correlation Between the Incidence of Major Side Effects
and the Protein A Serum Concentrations in Patients Treated
by Columns of Protein A with a Normal Fc Receptor Function

Protein A serum concentration (ng/ml)	No of sessions	No. of sessions with major side effects*
<10	152	3[a]**
10–50	31	6[b]
>50	12	5[c]

*The major side effects are the following: severe hypotension and tachycardia, chills, diarrhea, temperature elevation, or pulmonary rales. Side effects observed during perfusion of plasma from patients with melanoma and metastatic lesions (patients not mentioned in Table 2) have been included.

**The difference between (b) and (a), (c) and (a), (c) and (b) are statistically significant (p <0.01).

patients was perfused over columns of "denatured" protein A, i.e., protein A in which the Fc receptor function was chemically altered.

Leukocytic Histamine Release by Protein A

Some mild side effects (rash, urticaria) are reduced by intravenous injection of antihistaminic drugs. These side effects therefore might be the consequence of an interaction of solubilized protein A with circulating basophils and/or tissue mast cells. We therefore investigated in vitro the ability of protein A to release histamine from normal human leukocytes. Washed human leukocytes (2×10^7) were incubated for 30 minutes at 37°C in 2 ml Hepes containing Ca^{++}, Mg^{++}, and protein A (0.01 to 100 µg/ml). Histamine released from basophils was measured spectrofluorometrically in the supernatant. A significant histamine release ranging from 15% to 80% was observed with most of the healthy cell donors. The dose-effect relationship is a bell-shaped curve peaking at 1 µg/ml. Histamine release was found to be complete after 30-minute incubation. Protein A also causes a complement-dependent leukocytic histamine release. In summary, some allergic symptoms associated with plasma perfusion over protein A might be due to the release of mediators following the interaction of basophils with solubilized protein A.[63]

Interaction Between Protein A and Platelets

Staphylococci that possess protein A cause aggregation of human platelets in whole plasma accompanied by release of serotonin.[64] The interaction between staphylococci and platelets required the presence of cell wall-bound protein A and IgG with an intact Fc fragment.[64] Binding of protein A−IgG complexes to human platelet Fc receptors was paralleled by the release of serotonin. Such mechanisms are responsible for thromboembolic complications occurring at the sites of intravascular staphylococcal infections.[64] These side effects are highly improbable during the ex vivo plasma perfusion over protein A columns since:

(1) whole staphylococci are not used;
(2) the plasma separated by the macromolecular filter and perfused over the protein A columns contains less than 5,000 platelets per mm^3;
(3) no "infusion" of silica-bound protein A into the patient may occur if adequate filters are used;
(4) soluble protein A does not induce measurable platelet aggregation; however, soluble protein A can induce a serotonin release.[64]

A release of serotonin from platelets may have occurred in patient 8, who developed an acute respiratory distress syndrome associated with a marked thrombocytopenia[65] after the second session. A quinazoline derivative (Katanserin; R 41 468, Janssen Pharmaceutica, Belgium), selective blocker of 5-HT2 receptors,[66] has been shown to decrease both preload and afterload and to improve cardiac function in patients with heart failure, probably by a blockade of the vascular effects of serotonin released by aggregating platelets.[67] A significant, albeit transient, improvement of arterial blood gases was noted in our patient after intravenous injection of 10 mg of R 41 468.

Consumption of Complement Components

Arthus-like reactions can be produced in rabbits by interaction of protein A with human IgG.[1,68,69] Complexes between IgG and protein A activate the serum complement system in a manner that is entirely analogous to complement activation by antigen-antibody complexes.[51] Furthermore, the alternate complement pathway is also activated by addition of protein A to human serum.[51] We have noted a statistically significant reduction in the C3 proactivator, C3, C4 plasma levels, and in the total complement activity (CH50) immediately after perfusion of plasma over columns of "active" protein A; these values returned to their initial levels within 24 to 48 hours after the procedure. We have not been able to demonstrate any correlations between the consumption of complement components, the protein A serum concentrations, and the frequency or intensity of major side effects.

In summary, side effects of plasma perfusion over protein A are frequently observed. They are most probably the consequence of immune reactivity between protein A (and mainly its Fc receptor) and various blood components including: (1) circulation basophils and tissue mast cells; (2) circulating platelets; and (3) plasma complement components. Normal resting arterial blood gases and spirometry should also be required in patients presenting with lung metastatic lesions.[47,58]

Necrolytic Response

Two major mechanisms might be responsible for the tumoricidal response: (1) binding of IgG, of immune complexes and/or of "blocking factors" to protein A; (2) the appearance of some stimulatory activity in plasma after perfusion over protein A.

The first mechanism might be the consequence of either binding of Fc domains of IgG to protein A Fc receptors located inside the column, or "neutralization" of some circulating immune complexes by Fc receptors infused into the patient after leaking, as shown in vitro.[70] The second mechanism might be related to either a modification by the protein A columns of antigen-antibody ratios[41] or to some stimulatory properties of protein A infused into the patient after leaking.

Arguments in Favor of a "Column Effect"

The material eluted from the protein A column by 3M KSCN has been studied by immunological and biochemical methods.[71] The material eluted from "active" protein A columns contained 0.5 to 2 g of protein, while less than 0.2 g of protein was obtained from eluates from "Fc denatured" protein A columns. In the former eluates, 80% of the protein was composed of IgG and C3, whereas in the latter IgM, IgA, IgG, and C3 were detected in equal amounts.[71] The IgG eluted from "active" protein A columns strongly stained autologous adenocarcinoma cells, in contrast to IgG eluted from "Fc denatured" protein A columns.[58,71] These results can be correlated with the following data (Table 4):

(1) Prior to perfusion over protein A columns, tumor-associated antibodies were undetectable in the sera from the patient, and circulating immune complexes were generally detectable by the Clq binding assay or by the conglutinin assay;

(2) Within 4 to 24 hours after perfusion over "active" protein A columns, tumor-associated antibodies were observed in the same sera, and immune complex−plasma levels generally decreased;

Table 4
Indirect Immunofluorescence with
Serum Samples and Protein A Eluates[a]

Patient	Preperfusion samples	Postperfusion samples (4 hr)	Protein A eluates (1 mg protein/ml PBS)[b]
Group A			
2	—	+ +	+ + +
3	—	+ +	+ + +
Group B			
9	—	—	—
11	—	—	—

[a]The intensity of staining was evaluated by two independent observers on duplicate slides containing frozen sections of autologous tumor biopsies taken before starting the plasma perfusion over protein A, using a scale from — to + + +.
[b]PBS = phosphate-buffered saline.

(3) IgG and IgG-associated polypeptides (immune complexes?) were seen by immunoelectrophoresis and by SDS-polyacrylamide gel electrophoresis in eluates from "active" protein A columns;

(4) The material eluted from these columns significantly decreased the ^{3}H thymidine incorporation by normal human lymphocytes stimulated by phytohemagglutinin.

Taken together, the results obtained with protein A eluates strongly suggest that the "ex vivo" removal of circulating IgG and immune complexes from plasma of patients presenting with metastatic adenocarcinoma may elicit the tumoricidal response.

Arguments in Favor of a "Protein A Leaking Effect"

We have demonstrated by RIA that protein A is infused into some patients during perfusion of plasma over our columns of protein A. It seems reasonable, therefore, to suggest that this protein A-containing plasma might exhibit in vivo some stimulatory activity, as it has been shown for protein A in vitro, i.e., lymphocyte stimulation,[49,72,73] interferon production, and augmentation of natural killer cell activity.[50]

No parallelism has been found, however, among the protein A leaking, the protein A serum concentrations, and the tumoricidal response (Table 5). In normal rats, ^{125}I-labeled protein A (Fig. 4) accumulated in the spleen 2 hours after intravenous injection, particularly when the labeled protein A was mixed for 30 minutes at 37°C with autologous rat serum before injection.[74] At the same time protein A was detectable by RIA in the serum of some rats only.[74] The absence of parallelism between the protein A serum concentrations and the tumoricidal response must therefore be interpreted cautiously, since protein A might accumulate, as in rats, into the spleen of the patients after leaking. Further experiments must be conducted before assuming a beneficial effect of protein A leaking during plasma perfusion over protein A columns in human neoplasms.

SUMMARY AND FUTURE DEVELOPMENTS

Plasma from 11 patients presenting with metastatic adenocarcinoma was perfused ex vivo over columns of protein A covalently linked to crystalline silica. The plasma of eight patients (group A) was perfused over columns of "active"

Table 5
Relation Between the Protein A Serum Concentrations and the Tumoricidal Response in Patients Treated by Columns of Protein A with a Normal Fc Receptor Function

Patient	No. of sessions	No. of sessions with demonstrable protein A leaking	Protein A serum concentration** (ng/ml)
Responder (4)*	41	4	82 ± 15
Nonresponder (4)	33	3	74 ± 14

*The number in brackets represents the number of patients.
**Mean values ± 1 SD; the difference between the "responder" and "nonresponder" groups is not statistically significant.

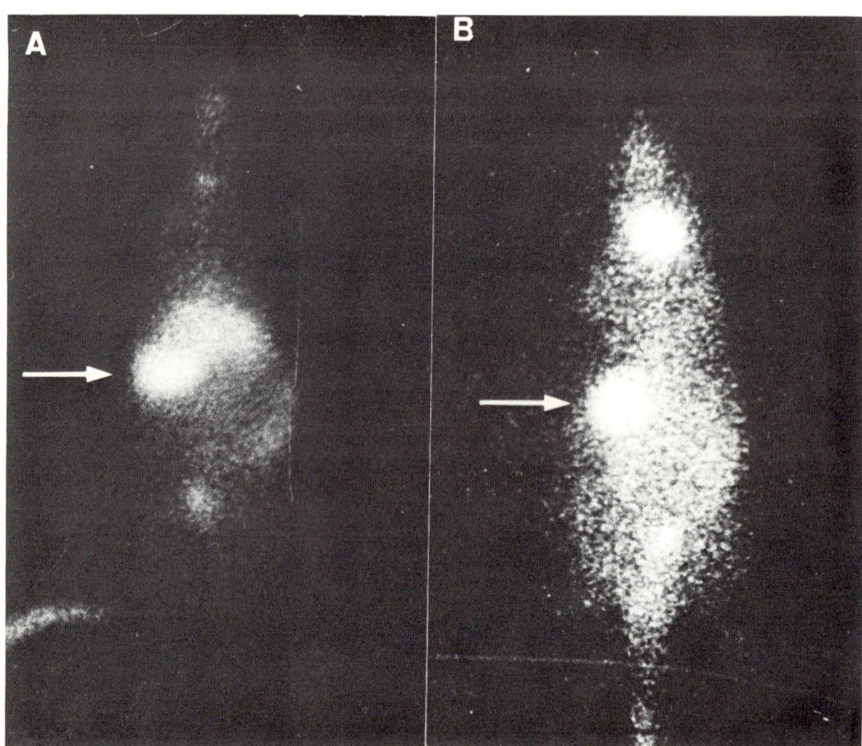

Figure 4: *Whole body scan of a rat after intravenous injection of* ^{125}I*-labeled protein A first mixed with autologous serum.* **(A)** *Two hours after injection; the spleen and the liver are particularly obvious (arrow).*
(B) *Twenty-four hours after injection; the radioactivity is concentrated into the spleen (arrow), whereas the liver is no longer obvious. The increase in thyroid radioactvity is probably related to free iodine.*

protein A, i.e., protein A exhibiting a normal Fc receptor-binding activity, and the plasma of three (group B) over columns of "denatured" protein A, i.e., protein A in which the Fc receptor-binding activity has been destroyed in order to determine whether the Fc receptor-binding activity of protein A was involved in the eventual tumoricidal response.

In group A, most patients developed acute, generally manageable, side effects (hypotension, chills, mild fever, leukocytosis, and pain in tumor sites) during (or immediately after) immunoadsorption procedures and four patients exhibited an objective partial regression after two to five sessions. In contrast, none of the patients from group B had any acute side effects and they showed evident progression of their disease during the procedure. Tumor cells in all stages of destruction and fibrosis were seen in biopsy specimens from patients in group A. However, focal areas of active tumor proliferation were always present in these specimens. These data show that ex vivo perfusion of plasma from patients with cancer over protein A columns induces a tumoricidal response in some instances, and that the Fc receptor function of protein A probably plays a major role in this necrolytic effect.

The mechanisms of the side effects remain unclear. We have shown that protein A leaking may occur during ex vivo perfusion of plasma over our protein A columns. Some undesirable side effects may be therefore the consequence of interactions between protein A infused into the patient and blood-formed elements, i.e., platelets and basophils. Furthermore, it is possible that some immune reactions between protein A and various components of plasma could release pharmacological agents, some of which might be responsible for side effects. The consumption of human complement components by complexes of IgG with protein A[51] as well as a significant decrease of C3 levels observed in the postperfusion plasma samples support this hypothesis.

The mechanisms responsible for the tumoricidal response also remain unknown. The absence of cellular infiltration in the tumors suggests that the tumoricidal effect may be mediated by humoral rather than cellular mechanisms. However, since protein A infused into the patient can accumulate in the spleen[74] and can induce interferon,[50] one may speculate whether protein A might not act, as interferon, by stimulating either some T cell functions[75] or natural killer cell activity.[50]

Whatever mechanisms are involved in the tumoricidal response, the results reported here and by others[47] appear sufficiently promising to warrant consideration in the management of such patients with poor prognosis. However, in order to delineate further the safety and efficacy of this immunoadsorption procedure, much more clinical and experimental work will be required. These studies must include:

(1) multicentric clinical trials in order to determine the indications, adverse effects, eventual therapeutic effectiveness, and long-term effects of this method in different types of cancer;

(2) multicentric clinical trials in order to test the efficacy, side effects, and protein A leaking of different types of protein A columns;

(3) experimental studies on the effects of protein A injection on tumor growth in animals with cancer;

(4) experimental studies on the mechanisms involved in the necrolytic response.

Acknowledgments

We are indebted to Mrs. Y. Pirard and A. Desoroux for their skillful technical assistance.

REFERENCES

1. Goudswaard J, van der Donk JA, Noordzij A, et al: Protein A reactivity of various mammalian immunoglobulins. *Scand J Immunol* 1978; 8:21.

2. Langone JJ, Boyle MDP, Borsos T: ^{125}I protein A: Applications to the quantitative determination of fluid phase and cell-bound IgG. *J Immunol Methods* 1977; 18:281.

3. Langone JJ: [^{125}I]protein A: A tracer for general use in immunoassay. *J Immunol Methods* 1978; 24:269.

4. Arend WP, Mannik M: The macrophage receptor for IgG: Number and affinity of binding sites. *J Immunol* 1973; 110:1455.

5. Segal DM, Hurwitz E: Binding of affinity cross-linked oligomers of IgG to cells bearing Fc receptors. *J Immunol* 1977; 118:1338.

6. McDougal JS, Redecha PB, Inman RD, et al: Binding of immunoglobulin G aggregates and immune complexes in human sera to staphylococci containing protein A. *J Clin Invest* 1979; 63:627.

7. Gold P, Freedman SO: Demonstration of tumor-specific antigens in human colonic carcinomata by immunological tolerance and absorption techniques. *J Exp Med* 1965; 121:439.

8. Morton DL, Malmgren RA: Human osteosarcomas: Immunologic evidence suggesting an associated infectious agent. *Science* 1968; 162:1279.

9. Morton DL, Malmgren RA, Holmes EC, et al: Demonstration of antibodies against human malignant melanoma by immunofluorescence. *Surgery* 1968; 64:233.

10. Old LJ, Boyse EA, Oettgen HF, et al: Precipitating antibody in human serum to an antigen present in cultured Burkitt's lymphoma cells. *Proc Natl Acad Sci USA* 1966; 56:1699.

11. Hellström IE, Hellström KE, Pierce GE, et al: Demonstration of cell-bound and humoral immunity against neuroblastoma cells. *Proc Natl Acad Sci USA* 1968; 60:1231.

12. Bubeník J, Perlmann P, Helmstein K, et al: Immune response to urinary bladder tumours in man. *Int J Cancer* 1970; 5:39.

13. Order SE, Porter M, Hellman S: Hodgkin's disease: Evidence for a tumor-associated antigen. *N Engl J Med* 1971; 285:471.

14. Hellström I, Hellström KE, Bill AH, et al: Studies on cellular immunity to human neuroblastoma cells. *Int J Cancer* 1970; 6:172.

15. Hellström I, Hellström KE, Shepard TH: Cell-mediated immunity against antigens common to human colonic carcinomas and fetal gut epithelium. *Int J Cancer* 1970; 6:346.

16. Nairn RC, Nind APP, Guli EPG, et al: Immunological reactivity in patients with carcinoma of colon. *Br Med J* 1971; 4:706.

17. O'Toole C, Perlmann P, Unsgaard B, et al: Cellular immunity to human urinary bladder carcinoma. II. Effect of surgery and preoperative irradiation. *Int J Cancer* 1972; 10:92.

18. O'Toole C, Perlmann P, Unsgaard B, et al: Cellular immunity to human urinary bladder carcinoma. I. Correlation to clinical stage and radiotherapy. *Int J Cancer* 1972; 10:77.

19. Fossati G, Canevari S, Della Porta G, et al: Cellular immunity to human breast carcinoma. *Int J Cancer* 1972; 10:391.

20. Hellström I, Hellström KE, Sjögren HO, et al: Demonstration of cell-mediated immunity to human neoplasms of various histological types. *Int J Cancer* 1971; 7:1.

21. Hellström I, Evans CA, Hellström KE: Cellular immunity and its serum-mediated inhibition in Shope-virus-induced rabbit papillomas. *Int J Cancer* 1969; 4:601.

22. Hellström I, Hellström KE: Colony inhibition studies on blocking and non-blocking serum effects on cellular immunity to Moloney sarcomas. *Int J Cancer* 1970; 5:195.

23. Sjögren HO, Bansal SC: Antigens in virally induced tumors. *Prog Allergy* 1971; 1:921.

24. Steele G Jr, Ankerst J, Sjögren HO: Alteration of in vitro anti-tumor activity of tumor-bearer sera by absorption with *Staphylococcus aureus*, Cowan I. *Int J Cancer* 1974; 14:83.

25. Hellström I, Hellström KE: Studies on cellular immunity and its serum-mediated inhibition in Moloney virus-induced mouse sarcomas. *Int J Cancer* 1969; 4:587.

26. O'Neill PA, Romsdahl MM: IgA as a blocking factor in human malignant melanoma. *Immunol Commun* 1974; 3:427.

27. Alexander P: Escape from immune destruction by the host through shedding of surface antigens: Is this a characteristic shared by malignant and embryonic cells? *Cancer Res* 1974; 34:2077.

28. Brown RJ: In vitro desensitization of sensitized murine lymphocytes by a serum factor (soluble antigen?). *Proc Natl Acad Sci USA* 1971; 68:1634.

29. Currie GA, Basham C: Serum-mediated inhibition of the immunological reactions of the patient to his own tumour: A possible role for circulating antigen. *Br J Cancer* 1972; 26:427.

30. Jose DG, Seshadri R: Circulating immune complexes in human neuroblastoma: Direct assay and role in blocking specific cellular immunity. *Int J Cancer* 1974; 13:824.

31. Sjögren HO, Hellström I, Bansal SC, et al: Suggestive evidence that the "blocking antibodies" of tumor-bearing individuals may be antigen-antibody complexes. *Proc Natl Acad Sci USA* 1971; 68:1372.

32. Baldwin RW, Price MR, Robins RA: Blocking of lymphocyte-mediated cytotoxicity for rat hepatoma cells by tumour-specific antigen-antibody complexes. *Nature New Biol* 1972; 238:185.

33. Hellström I, Warner GA, Hellström KE, et al: Sequential studies on cell-mediated tumor immunity and blocking serum activity in ten patients with malignant melanoma. *Int J Cancer* 1973; 11:280.

34. Baldwin RW, Embleton JM, Price MR: Inhibition of lymphocyte cytotoxicity for human colon carcinoma by treatment with solubilized tumor membrane fractions. *Int J Cancer* 1973; 12:84.

35. Currie G: The role of circulating antigen as an inhibitor of tumor immunity in man. *Br J Cancer* 1973; 28 (Suppl I):153.

36. Browne O, Bell J, Holland PDJ, et al: Plasmapheresis and immunostimulation. *Lancet* 1976; 2:96 (letter to the editor).

37. Hersey P, Isbister J, Edwards A, et al: Antibody-dependent, cell-mediated cytotoxicity against melanoma cells induced by plasmapheresis. *Lancet* 1976; 1:825.

38. Isbister WH, Noonan FP, Halliday WJ, et al: Human thoracic duct cannulation: Manipulation of tumor-specific blocking factors in a patient with malignant melanoma. *Cancer* 1975; 35:1465.

39. Israel L, Edelstein R, Mannoni P, et al: Plasmapheresis in patients with disseminated cancer: Clinical results and correlation with changes in serum protein. The concept of "nonspecific blocking factors." *Cancer* 1977; 40:3146.

40. Bansal SC, Bansal BR, Rhoads JE Jr, et al: Ex vivo removal of mammalian immunoglobulin G: Method and immunological alterations. *Int J Artif Organs* 1978; 1:94.

41. Bansal SC, Bansal BR, Thomas HL, et al: *Ex vivo* removal of serum IgG in a patient with colon carcinoma: Some biochemical, immunological, and histological observations. *Cancer* 1978; 42:1.

42. Jones FR, Yoshida LH, Ladiges WC, et al: Treatment of feline leukemia and reversal of FeLV by ex vivo removal of IgG: A preliminary report. *Cancer* 1980; 46:675.

43. Jonsson S, Huberg B, Hakansson L, et al: Immunological tumor treatment with extracorporeal affinity chromatography of plasma using protein A—Sepharose. Molecular, biologic background and animal experiments. *Abstracts from the Annual Meeting of the Swedish Medical Society*, Stockholm, December 5-8, 1979.

44. Terman DS, Yamamoto T, Mattioli M, et al: Extensive necrosis of spontaneous canine mammary adenocarcinoma after extracorporeal perfusion over *Staphylococcus aureus* Cowans I. I. Description of acute tumoricidal response: Morphologic, histologic, immunohistochemical, immunologic, and serologic findings. *J Immunol* 1980; 124:795.

45. Lindholm L, Braide I, Persson B, et al: Preliminary experience with a staphylococcus protein A containing extracorporeal adsorption system for the treatment of advanced cancer. *Abstracts from the Annual Meeting of the Swedish Medical Society*, Stockholm, December 5-8, 1979.

46. Ray PK, Besa E, Idiculla A, et al: Efficient removal of abnormal immunoglobulin G from the plasma of a multiple myeloma patient: Description of a new method for treatment of the hyperviscosity syndrome. *Cancer* 1980; 45:2633.

47. Terman DS, Young JB, Schearer WT, et al: Preliminary observations of the effects on breast adenocarcinoma of plasma perfused over immobilized protein A. *N Engl J Med* 1981; 305:1195.

48. Hellström I, Hellström KE, Evans CA, et al: Serum-mediated protection of neoplastic cells from inhibition by lymphocytes immune to their tumor-specific antigens. *Proc Natl Acad Sci USA* 1969; 62:362.

49. Sakane T, Green I: Protein A from *Staphylococcus aureus*: A mitogen for human T lymphocytes and B lymphocytes but not L lymphocytes. *J Immunol* 1978; 120:302.

50. Ratliff TL, McCool RE, Catalona WJ: Interferon induction and augmentation of natural-killer activity by staphylococcus protein A. *Cell Immunol* 1981; 57:1.

51. Stålenheim G, Götze O, Cooper NR, et al: Consumption of human complement components by complexes of IgG with protein A of *Staphylococcus aureus*. *Immunochemistry* 1973; 10:501.

52. Sjöquist J, Meloun B, Hjelm H: Protein A isolated from *Staphylococcus aureus* after digestion with lysostaphin. *Eur J Biochem* 1972; 29:572.

53. Hunt J, Kinet JP, Foidart JB, et al: Measurement of protein A in biological fluids by competitive inhibition solid phase radioimmunoassay using F(ab')$_2$ fragments of anti-protein A antibody. *J Immunol Methods* (submitted for publication).

54. Sjöholm I: Protein A from *Staphylococcus aureus*: Spectropolarimetric and spectrophotometric studies. *Eur J Biochem* 1975; 51:55.

55. Weetall HH: Covalent coupling methods for inorganic support materials. *Methods Enzymol* 1976; 44:134.

56. Sieberth HG, Glöckner W, Hirsch HH, et al: Plasma-separation mit Hilfe von Membranen: Untersuchungen am Menschen. *Klin Wochenschr* 1980; 58:551.

57. Bensinger WI, Baker DA, Buckner CD, et al: Immunoadsorption for removal of A and B blood-group antibodies. *N Engl J Med* 1981; 304:160.
58. Kinet JP, Bensinger WI, Balland N, et al: Tumoricidal response in human neoplasms after ex vivo perfusion of plasma over protein A columns. Report of eleven cases and evidence for a major role of the Fc-receptor function of protein A in the necrolytic effect. *J Clin Apheresis* (submitted for publication).
59. Bensinger WI, Kinet JP, Hennen G, et al: Plasma perfused over immobilized protein A for breast cancer. *N Engl J Med* 1982; 306:935 (letter to the editor).
60. Hayward JL, Carbone PP, Heuson J-C, et al: Assessment of response to therapy in advanced breast cancer. *Cancer* 1977; 39:1289.
61. Hunt JS, Macdonald PR, McGiven AR: Characterisation of human glomerular basement membrane antigenic fractions isolated by affinity chromatography utilising anti-glomerular basement membrane auto-antibodies. *Biochem Biophys Res Commun* 1982; 104:1025.
62. Bolton AE, Hunter WM: The labeling of proteins to high specific radioactivities by conjugation to a ^{125}I-containing acylating agent: Application to the radioimmunoassay. *Biochem J* 1973; 133:529.
63. Radermecker M, Kinet JP, Hennen G, et al: Protein A induced leukocytic histamine release. *J Allergy Clin Immunol* 1983; 71:134.
64. Hawiger J, Steckley S, Hammond D, et al: Staphylococci-induced human platelet injury-mediated by protein A and immunoglobulin G Fc fragment receptor. *J Clin Invest* 1979; 64:931.
65. Schneider RC, Zapol WM, Carvalho AC: Platelet consumption and sequestration in severe acute respiratory failure. *Am Rev Respir Dis* 1980; 122:445.
66. Leysen JE, Awouters F, Kennis L, et al: Receptor binding profile of R 41 468, a novel antagonist at 5-HT$_2$ receptors. *Life Sci* 1981; 28:1015.
67. Demoulin J-C, Bertholet M, Soumagne D, et al: 5-HT$_2$-receptor blockade in the treatment of heart failure. *Lancet* 1981; 1:1186.
68. Goding JW: Use of staphylococcal protein A as an immunological reagent. *J Immunol Methods* 1978; 20:241.
69. Gustafson GT, Sjöquist J, Stålenheim G: "Protein A" from *Staphylococcus aureus*. II. Arthus-like reaction produced in rabbits by interaction of protein A and human γ-globulin. *J Immunol* 1967; 98:1178.
70. Cowan FM, Klein DL, Armstrong GR, et al: Neutralization of immune complex inhibition of antibody-dependent cellular cytotoxicity in vitro by *Staphylococcus aureus* protein A. *Biomedicine* 1979; 30:23.
71. Kinet JP, Bensinger B, Hennen G, et al: Treatment of breast cancer by ex vivo immunoadsorption of plasma on protein A columns: Correlations between the immunochemical analysis of protein A eluates and the therapeutic effectiveness. *Eur J Clin Invest* 1982; 12:19 (abstract).
72. Berman MA, Spiegelberg HL, Weigle WO: Lymphocyte stimulation with Fc fragments. I. Class, subclass, and domain of active fragments. *J Immunol* 1979; 122:89.

73. Berman MA, Weigle WO: B-lymphocyte activation by the Fc region of IgG. *J Exp Med* 1977; 146:241.
74. Hoyoux C, Kinet JP, Foidart JB, et al: Studies on the kinetic of blood and tissue distribution of ^{125}I-protein A intravenously injected into normal rats. *Clin Exp Immunol* (submitted for publication).
75. Krown SE, Burk M, Kirkwood JM, et al: Human leukocyte interferon (HuLeIF) in malignant melanoma (MM): Preliminary report of the American Cancer Society Clinical Trial. *Proc Am Assoc Cancer Res Am Soc Clin Oncol* 1981; 22:158.

CHAPTER 7

Chemical Precipitation and Removal of IgG and Immune Complexes

David H. Bing, Ph.D.

INTRODUCTION

The method of volume replacement in therapeutic plasma exchange has received major attention.[1] Two particular concerns have been: (1) the undetermined consequences that result from continually removing normal plasma protein components, and (2) the nature of the replacement fluid. When normal plasma is used to replace removed volumes, there is the risk of hepatitis and hypersensitivity reactions. While normal serum albumin is a safer replacement fluid, its use represents a significant portion of the cost of the procedure. If methods could be devised that allowed for the retention of normal plasma proteins in a physiologically safe form in the plasma being removed, the patient's own plasma could be used as the replacement fluid. While attention has been devoted to developing selective absorbents for removal of un-

This research was supported by NIH grant HL 24856 and by a grant from the Muscular Dystrophy Association.

139

wanted plasma proteins during plasmapheresis,[2,3] little work has been done to determine if separation methods which are based on the physicochemical properties of the proteins might yield safe plasma protein derivatives, such that at the least, the patient's own albumin could be used as the replacement fluid.

Our laboratory has been studying two physicochemical fractionation methods that permit retention of the albumin and other plasma proteins while removing IgM and IgG and immune complexes. Both methods could be used in conjunction with plasmapheresis. The first is isoelectric precipitation of plasma by rapid deionization. This method leads to removal of IgM and immune complexes, both of which are relatively insoluble at low ionic strengths. The other technique is selective precipitation of IgG and immune complexes by zinc diglycinate, a metal ion complex which was developed as an alternative to the cold ethanol technique for the preparation of immunoglobulin concentrates.

Isoelectric precipitation is based on the fact that proteins are most insoluble at their isoionic points. In aqueous solution, proteins are polyionic and bind water as well as ions. Experimentally determined densities of proteins usually assume that bound water is the same as bulk water, but x-ray crystallographic analyses of proteins have shown this is not true.[4] Richards and Richmond have suggested, in fact, that water bound to proteins in deep grooves (e.g., active sites) has a substantially higher fugacity than bulk water, can be more easily displaced, and thus contributes significantly to ligand binding energy.[5] When present, ions of neutral salts (e.g., those that do not change pH in an aqueous solution) shield charged groups of proteins and thus affect the electrostatic interactions between such groups in proteins. The effect occurs at low salt concentrations; physiological salt concentrations and typical ionic strengths used in vitro (.05 to 0.2 and even as low as .01) are sufficient to minimize such interactions. Neutral salts can promote stable protein conformations (helix, native conformation)[6] and the stabilizing effects follow the classical Hofmeister series that ranks the effectiveness of ions in the salting-out of the protein.[7] In contrast, increased solubility leads to decreased stability; denaturation by 6 M urea or 6 M guanidinium ions is an example of maximal solubility and loss of structural integrity.

The fractional precipitation of plasma proteins by controlling ionic strength and pH is based on the protein solubility properties just detailed and reflects the balance between hydrophilic and hydrophobic groups on the molecule. The relationship between the solubility of a protein and ionic strength is expressed by the relationship[8]:

$$\text{Log } S = \beta - K_s \cdot I \tag{1}$$

in which S = solubility, I is ionic strength, K_s is the salting-out coefficient, and β is a pH and temperature parameter. The pH dependency of salting-out reflects the fact that a protein has maximal insolubility in electrolyte-free medium at its isoionic point. This phenomenon is known as isoelectric precipitation; it follows, and has been observed, that a protein can be quite soluble in low ionic strength medium provided it is not at a pH close to its isoionic point.[7,9] Salting-out as well as salting-in of a protein is complicated by the presence of one or more proteins; under such conditions there are no sharp transitions in solubility as the concentration of the proteins and/or ions change. Use of these principles is the basis of salting-out, isoelectric precipitation, and salting-in as standard methodology in the fractionation of plasma proteins. Salts such as $(NH_4)_2SO_4$ and Na_2SO_4 are the most commonly used, but polyanions (phosphate, polyacrylic acid) have been used as well. Fractional protein precipitation by removal of electrolytes via dialysis against water or diluted buffers leads to the formation of an insoluble fraction or euglobulin.[9] Dialysis of serum against water results in a slow decrease in pH and leads to the formation of coacervates, particularly when the pH reaches a value lying between isoelectric points of different proteins in the mixture. Electrodialysis is superior in the sense that the imposition of a direct current (DC) to increase ion flux across the membrane results in faster and more complete removal of electrolytes.[10]

We have applied the use of electrodialysis across ion exchange membranes with rapid mixing to the study of the behavior of the plasma proteins. This technology has evolved from the use of an electrodialysis process to remove salt from cheese whey.[11-13] Electrodialysis, as we have developed it, is a process in which ions are transferred from one solution through selectively permeable membranes into another solution by imposition of a DC electrical potential. The membranes contain covalently bound ion exchange groups which have either a positive or negative fixed electrical charge. Positively charged anion membranes will pass anions and repel cations. Negatively charged cation membranes will pass cations and repel anions. When a voltage is applied across an ion exchange membrane in an electrolyte, the resulting electrical current is passed only by the ions that have a charge opposite to that of the membrane. For this reason, "stacks" of cation and anion permeable membranes alternately will concentrate or dilute electrolytes in the solutions in contact with them (see Fig.1).[11] Two cation membranes and one anion membrane constitute what is called one cell pair.

It should be emphasized that in contrast to dialysis, there is movement of ions rather than bulk electrolytes. This is different from dialysis in which differences in solute concentration across semipermeable membranes determines the final concentration of ions. In electrodialysis, the degree to which a solution is desalted is proportional to the electrical current which in turn is limited by electrical resistance of the stack components. Furthermore, electrodialysis is a flow-through system that contains no filters or matrices which can cause changes in pressure or flow rates. The molecular weight exclusion limit of the membrane, as we have developed it, is 10,000, so there is no significant protein loss. The electrodialysis across ion exchange membranes (EDIM) with rapid mixing results in low membrane potential (5 to 10 volts),

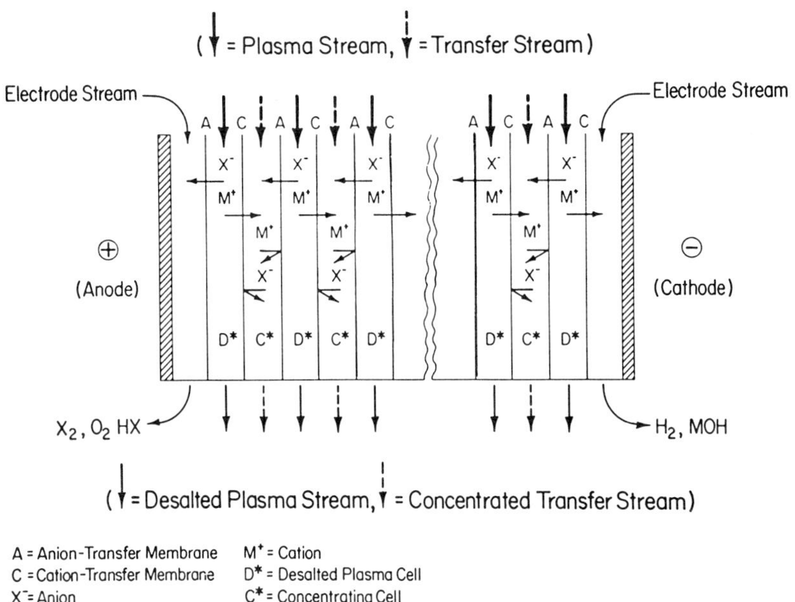

Figure 1: *Schematic representation of electrodialysis apparatus used in deionization of plasma.*

causes no large pH shifts due to accumulation of ions, and gives exact control over the deionization process.

Selective precipitation of immunoglobulins by zinc ions was originally described by Cohn,[14] but it was Isliker and Antoniades who showed definitively that Zn^{++} could deplete isoagglutinin levels in serum.[15] Subsequently, Surgenor and co-workers[16,17] devised a fractionation scheme called Method 12, which used metal salts (barium sulfate and zinc diglycinate) to precipitate coagulation factors and immunoglobulins and which yielded a fraction called stable plasma protein solution (SPPS) that could be used in place of 25% albumin as a plasma extender.[18] Schultze and Heremans have reviewed precipitation of plasma proteins by metal salts and have suggested that the mechanism of selective zinc precipitation of IgG is via interaction with the side chain of histidine. The selectivity of zinc is presumably based on its ability to complex with imidazole ring of histidines.[7]

METHODOLOGY

Proteins

Fresh frozen CPD plasma was obtained from the American Red Cross Blood Services, Northeast Region, Boston, Mass. Plasma, collected during therapeutic plasmapheresis, was kindly provided by Dr. Jettie Hunt, Massachusetts General Hospital, Boston, Mass. Diagnoses were provided by attending physicians. Plasmas were maintained frozen at $-20°C$ until used. They were thawed strictly at $15°C$ in a circulating water bath and used without further treatment. Serum immune globulin, tetanus toxoid, and tetanus immune globulin (TIG) were generously provided by Dr. Jeanne Leszczynski, Biologic Laboratories, State Laboratory Institute, Jamaica Plain, Mass. Clq was purified as described by Bing et al.[19] Antiserum to Clq was obtained from Atlantic Antibodies, Westbrook, Maine.

Assays

Immune complex assays were done with the solid-phase Clq binding assay as described by Wehler et al.[20] as modified by Bourke et al.[21] in which ^{125}I-labeled protein A was used to detect bound immunoglobulin. Passive hemagglutination assays used glutaraldehyde-fixed sheep erythrocytes sensitized with tetanus toxoid as described by Bing et al.[22] Heat-aggregated IgG was prepared as previously described[20] by heating 1% solutions at $63°C$ for 10 minutes. Both IgG and heat-aggregated IgG were iodinated by the chloramine T method[23] to a specific activity of 15 to 20 μCi/mg. Immuno-chemical analysis of plasma protein levels was performed by the turbidometric technique with the Technicon Autoanalyzer as described by Ritchie[24,25] in the Blood Grouping Laboratory at the Center for Blood Research.

Serum protein analyses results have been presented in terms of g/dl, mg/dl, and % of normal (%nl). The absolute values are related to values determined in 200 normal individuals (approximately 60% Caucasian and 40% black) randomly selected from the northeast region of the United States. Where results are expressed as %nl, these are based on the determination of protein concentrations in 200 samples as compared to a pool of normal serum in which every contributor was proven to be normal in all proteins. Thus, normal level is expressed as 100%nl and decreases or elevation would be less than or greater than 100%nl. In the case of immune complex determinations, levels have been determined similarly, only comparison has been made to the mean of 10 proven normal samples collected and treated identically to the sample in question and assayed simultaneously with the particular sample. Ranges are expressed as 2 standard deviations. The value is considered abnormal if it is outside that range.

Factor X and factor II assays used reagents purchased from Sigma Chemical Co., St. Louis, Mo., and values were normalized to the lyophilized standard provided. Total protein was measured with an American Optical refractometer with directions supplied by the manufacturer. Albumin was also measured chemically with bromcresol green (Fisher Scientific). Fibrinogen was measured as clottable fibrinogen with reference to a standard of purified fibrinogen.[26] Factor VIII-C was measured as described by Hardisty and Macpherson[27] and compared to a pool of normal fresh frozen plasma which was predetermined to have normal factor VIII-C activity. This was generously donated by Dr. Barbara C. Furie, Tufts-New England Medical Center, Boston, Mass. Cl was measured as described by Rapp and Borsos[28] modified by Bing et al.[19] in which volumes are reduced by one-fifth to conserve reagents.

Clq was determined by radial immunodiffusion[29] in 0.8% agarose containing 3.3% antiserum and buffered in 75 mM NaCl, 25 mM Tris, 20 mM glycine, 0.5 mM EDTA, pH 7.4. The precipitin rings were developed 48 hours at room temperature prior to washing and staining with 1.0% amidoblack in 5% glacial acetic acid. A standard curve was constructed with purified Clq.

Crossed immunoelectrophoresis was performed as described by Laurell[30] with monospecific goat anti-C3 in the second phase. To activate the C3 in the plasma, heat-aggregated IgG (1 mg) or zymosan (2 mg) was added in 50 μl to 200 μl of plasma and incubated at 37°C for 1 hour. A control consisted of 200 μl of plasma plus 50 μl of 150 mM NaC1/10 mM $NaPO_4$, pH 7.2, treated identically. Gels were washed free of unprecipitated protein and stained with Coomassie Blue R250, the stain patterns were transferred to graph paper by xerography, and the areas under the curves determined by digitization on the PROPHET computer with the programs DIGITIZE and INTEGRAL.[31]

Electrodialysis cells with permeable selective cation and anion membranes were obtained from Ionics, Inc., Watertown, Mass. Two types of electrodialysis cells were used. The Dial-a-Cell apparatus was used equipped with three to four cell pairs with a total exposed membrane area per cell pair of 6.82 cm^2. The total volume in the electrodialysis cell at any time was 2–5 ml. Cole Parmer pumps equipped with 7014 Masterflex pump heads and 0.0625 × 0.13 inch (i.d.) Tygon tubing were used to circulate the plasma and electrolyte solutions. An LKB 2103 power supply was the power source. Temperature was maintained at 0° to 5°C by immersion of the electrodialysis cell in an ice bath. Flow rates were maintained at 100 ml/min/cell. All solutions and plasma were maintained at 15°C by immersion in a circulating 15°C water bath. Temperature of the solutions were monitored and did not rise above 15°C. Conductivity and pH were measured with a Radiometer model CDM2 conductivity meter and model 26 pH meter, respectively, with samples equilibrated to 0°C. Samples were collected manually at designated times. For the larger volumes of plasma, the Ionics Medimat was used equipped with one cell pair (see Fig. 1). The size of the cell is larger (9×10 inches) and the total membrane area available per cell pair was 220 cm^2. The Medimat contained a SCR-10 model 300-5 power source (CE/M Measurements, Inc., Boston, Mass.) and an Extech model 480 pH/conductivity probe (Extech, Boston, Mass.). Circulation was accomplished with a built-in Cole Parmer pump with two 7018 and two 7017 heads and $\frac{3}{8}$ inch Tygon tubing. Cooling was accomplished by immersion of electrolytes and plasma in an ice bath. The cell was maintained at 4°C by an FTS Recirculating Cooler, Tarrytown, N.Y., by circulating coolant at about 100 ml/min through water cooling plates which were included as end-spacers in the cell. With the Medimat, samples were collected by means of an LKB Redirack fraction collector. Percent desalting was determined based on the reduction of the conductivity of the plasma. The cell was assembled and tested for leaks by circulating cold water and then 0.15 M NaCl. In order to minimize dilution before introducing the plasma, electrode buffer, or concentrate solutions, excess liquid was removed by pumping out excess fluid. In spite of this there was always an initial 2–5% dilution of the plasma sample. However, during electrodialysis there was always a 5–7% reduction in volume due to water transfer during desalting, and this balanced out the initial dilution.

The electrodialysis was conducted at a constant i/N of 500 A/cm^2/g equiv/liters. As the salt content of plasma was depleted, the current was adjusted manually to a lower value to maintain this value. The extent to which current had to be adjusted to maintain constant i/N was determined from standard curves of current vs. conductivity constructed based on total membrane area and flow rates of solutions passing through the electrodialysis cell. The concentrate stream (i.e., the stream receiving the ions being removed) was 0.25% NaCl. The electrode stream was 2.5% Na citrate adjusted as indicated.

The method described by Pennell[18] was used to prepare zinc diglycinate and precipitate immunoglobulins.

RESULTS

Electrodialysis of Normal Plasma

When normal CPD plasma was thawed at 15°C and subjected to electrodialysis, there was a rapid removal of salt as indicated by a drop in the conductivity of the circulating plasma. Typically, the starting conductivity of the plasma was 8,500–11,000 μS. At the end of an electrodialysis procedure, the conductivity was 800–1,000 μS. Results from a typical EDIM procedure of 350 ml of plasma performed at i/N of 500 in a 220 cm^2 stack (Medimat) are presented in Figure 2. During

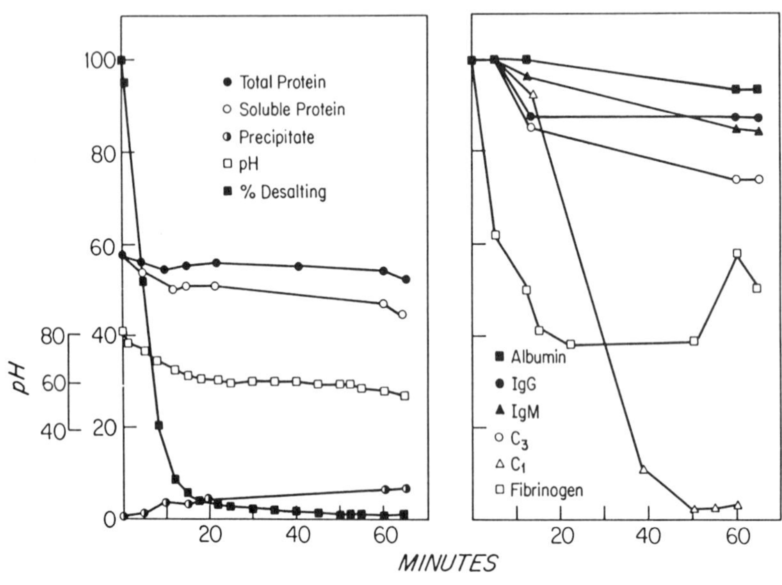

Figure 2: *Electrodialysis of normal CPD plasma. Electrodialysis was done in the Medimat with two cell pairs at a flow rate of 150 ml/min/channel. Desalting total plasma protein and soluble protein values are expressed as percent of original value in plasma. Precipitated protein is in mg/ml based on resolubilization to original sample volume.*

the electrodialysis the plasma circulated freely through the cell and no clotting occurred. The process was rapid (80% deionization occurred in less that 5−7 minutes). At selected points, aliquots of deionized plasma were removed and centrifuged at 2,000 × g to remove a precipitate which formed during deionization. The precipitate was resolubilized in 0.5 M NaCl to the original plasma volume and analyzed for total protein along with the clear supernatants by determining the absorbency at 280 nM. The concentration of the precipitate was 45 mg/ml at the point where 95% deionization occurred. Measurement of total serum protein by A_{280} in the supernatant and precipitate indicated there was very little loss of total protein during the electrodialysis. This was verified by the failure to elute any protein (A_{280} ≤ 0.05/ml) from the membranes by 0.1 N NaOH after flushing with 0.15 M NaCl at the end of the procedure. Albumin, IgG, IgM, C3, Cl, and fibrinogen analyses were performed only on the supernatants. There was little loss of albumin, about 10% loss of IgG and IgM, 50% depletion of clottable fibrinogen, and complete removal of Cl activity at about 97% deionization. Because the process with the 220 cm^2 apparatus was so rapid, a smaller system (Dial-a-Cell) was used to examine the process under slower (60−70 min) conditions of deionization of smaller plasma (100 ml) samples. During EDIM in the 6.82 cm^2 cell, the pH gradually fell throughout the deionization process and the formation of a precipitate was detectable at a point where there was 50% deionization of the plasma. As in the larger EDIM cell, concomitant with the salt removal and lowering of the pH was the appearance of a fine white precipitate. The precipitate was collected at indicated points (Fig. 3) by centrifugation at 2,000 × g and resolubilized to original plasma volume in 0.5 M NaCl. To determine the dynamics of precipitate formation, this amount of protein precipitated at each point was expressed as a fraction of the total protein precipitated at 99% desalting. The inflection points at pH 6.0 and pH 6.5 were reminiscent of the euglobulin fractionation of plasma reported at these pHs by Sandor[32] and in which the three separate euglobulin fractions termed I, II, and III were formed

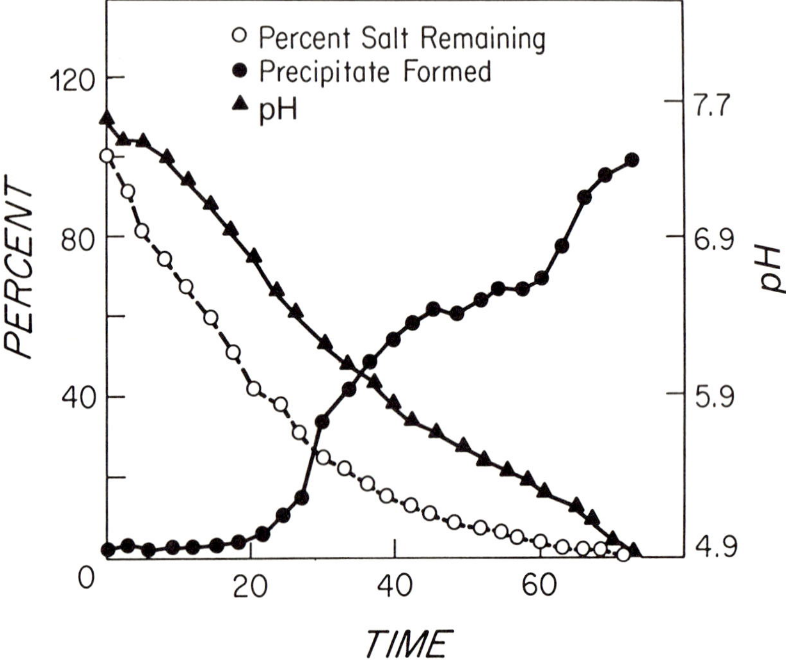

Figure 3: *Precipitation of proteins by EDIM as a function of decreasing ionic strength (○) and pH (▲). Several samples were subjected to serum protein analyses (see Table 1). Four cell pairs were used and the Dial-a-Cell was used (see* Methodology *section). Fraction of protein precipitated (●) was expressed as described in* Results. *Desalting (○) is expressed as percent of initial conductivity.*

(see Fig. 4). In the case of EDIM with the Dial-a-Cell, the final amount of precipitates formed was 35 mg/ml, but this is an underestimate, as the precipitates formed between 90% and 99% desalting (pH 6.0 to 5.0) could not be totally resolubilized in 0.5 M NaCl. For this reason the total recovery of protein at 99% desalting was only around 85%, and this fact is reflected by the generally overall lower values for all proteins analyzed (see below). The drop in pH and ionic strength simultaneously with the increase in precipitate formation suggested that isoelectric precipitation was occurring, in which case the protein, rather than the electrolyte solutions, controlled the final pH of the plasma stream as the deionization procedure

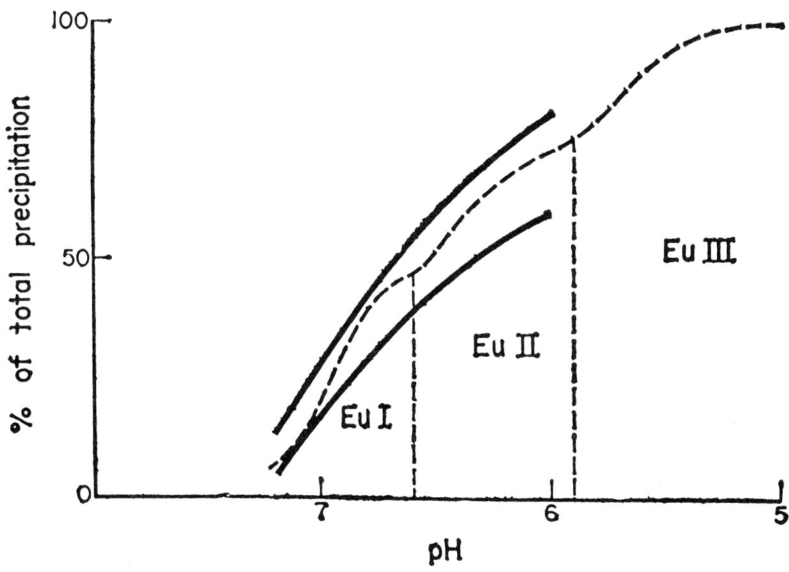

Figure 4: *Euglobulin fractionation of plasma at constant ionic strength (.005) and decreasing pH (from Sandor).*[32]

occurred. This was tested in an experiment in which 1% IgG solution and 4% and 0.4% albumin solutions adjusted to pH 7.0 in 0.15 M NaCl were deionized by EDIM. At the point where deionization proceeded from 90% to 99% desalting, the pH fell to 5.2 for albumin (pI of albumin $\simeq$ 4.8) or rose to 7.2 for IgG (pI = 6.8 to 7.2). If isoelectric precipitation was really occurring as described by Sandor (Fig. 4) during electrodialysis (Fig. 3), then protein separation should occur at the inflection points observed in protein precipitation when the 2.5% Na citrate pH 5.5 was used as the electrode stream (see Fig. 3). Plasma protein analyses of supernatant samples at various points in the desalting process by EDIM under these conditions are presented in Table 1; the euglobulin terminology of Sandor was used to categorize the fractions formed. Euglobulin I (pH 7.7 to pH 6.5) contained minor amounts of all of the plasma proteins measured. Clq, fibrinogen, C3, and C4 increased dramatically in euglobulin II (pH 6.5 to 6.0).

Table 1
Electrodialysis of Normal CPD Plasma with the Electrode Stream of pH 5.5
and Analysis of Supernatant Plasma Proteins

Parameter	Euglobulin I								Euglobulin II				Euglobulin III		
pH	7.69	7.54	7.50	7.39	7.26	7.09	6.90	6.74	6.24	6.11	6.00	5.60	5.46	5.23	4.93
% Desalting	0	9.7	18.5	26.1	32.6	41.3	49.5	59.7	75.5	79.8	72.1	90.0	94.7	97.6	99.5
% Precipitate	1.7	2.6	1.9	2.3	2.7	2.6	3.7	4.6	33.9	41.7	49.1	62.6	67.6	77.8	100.0
Total Protein g/dl	6.10	5.55	5.55	5.55	5.40	5.30	5.30	5.20	4.65	4.70	4.60	4.40	4.10	4.00	4.00
Albumin g/dl	3.10	3.00	3.00	3.00	3.00	3.10	3.10	3.10	3.00	3.05	3.10	3.10	3.10	3.10	3.20
Orosomucoid mg/dl	115	110	110	110	110	110	110	110	110	105	105	92	80	55	20
Haptoglobin mg/dl	126	124	124	124	126	124	124	123	120	121	120	105	105	92	72
Transferrin mg/dl	260	250	245	245	240	255	255	255	250	255	255	250	255	255	255
α_1-Antitrypsin (%nl)	55	55	50	50	55	50	55	55	55	55	55	55	22	1	1
Cl INH (%nl)	98	87	87	87	87	90	84	87	84	80	75	47	25	12	12
C3 mg/dl	98	98	102	98	98	104	84	84	60	40	30	12	10	10	6
C4 (%nl)	85	85	85	85	85	85	85	72	72	62	50	33	20	10	10
β-Lipoprotein mg/dl	52	50	50	52	50	48	50	50	43	47	47	42	42	38	34
IgM mg/dl	88	88	88	88	88	—	78	70	—	62	—	39	67	67	140
IgG mg/dl	850	880	880	880	880	880	870	870	810	810	790	750	740	750	750
IgA mg/dl	175	175	175	170	165	170	170	170	170	140	145	75	170	170	170
Clq (% initial value)	100	100	97.3	97.3	97.3	94.6	94.6	91.9	76.8	71.9	ND	ND	ND	ND	ND
Fibrinogen	100	—	100	—	94.8	—	92.3	77.6	—	46.2	34.1	31.1	34.1	31.0	29.4
Factor X	100	—	—	—	—	—	—	—	—	—	—	—	—	—	98
Factor VIII	100	—	—	—	—	—	—	—	—	—	—	—	—	—	95
Cl	100	—	—	—	—	—	—	—	—	—	—	—	—	—	ND

ND = none detected

Haptoglobin, orosomucoid, Cl, INH, α_1-antitrypsin, and IgM were found in euglobulin III (pH 6.0 to 5.0). IgG and β-lipoprotein appeared to be only partially depleted, and there was virtually no change in albumin, transferrin, and IgA. This fractionation was further evident in the examination of the color of the precipitates. Euglobulin I was white, euglobulin II blue, and euglobulin III yellow, exactly as reported by Sandor in his experiments.[32] It should be emphasized, however, that in the small scale as well as in the large scale experiment there was high retention of the albumin at 99% desalting of the plasma.

Because of the behavior of Cl and Clq under these conditions, and based on the known affinity of Clq for antigen-antibody aggregates, we decided to determine whether immune complexes might also be removed. This was examined by adding ^{125}I-labeled heat-aggregated IgG to normal plasma and then subjecting the mixture to electrodialysis. Precipitated protein was separated by centrifugation at 2,000 × g for 10 minutes, and aliquots of supernatants were used to determine remaining radiolabeled proteins. These data are presented in Figure 5. Electrodialysis of heat-aggregated ^{125}I-IgG with the electrode stream at pH 5.5 resulted in 80% removal of the labeled protein at 90−95% deionization. In all cases, however, as with the plasma proteins, the greatest change in levels of aggregated IgG occurred between 80% and 95% desalting.

Precipitation of Plasma Proteins from Normal Plasma with Zinc Diglycinate

Zinc diglycinate has previously been used at a final concentration of 50 mM to prepare immunoglobulin concentrates.[17,18] The effect of this and other concentrations of zinc diglycinate on selected plasma proteins was first examined. The results of this experiment (Fig. 6) confirmed the previous studies[15−18] that 50 mM zinc diglycinate was optimal to separate albumin

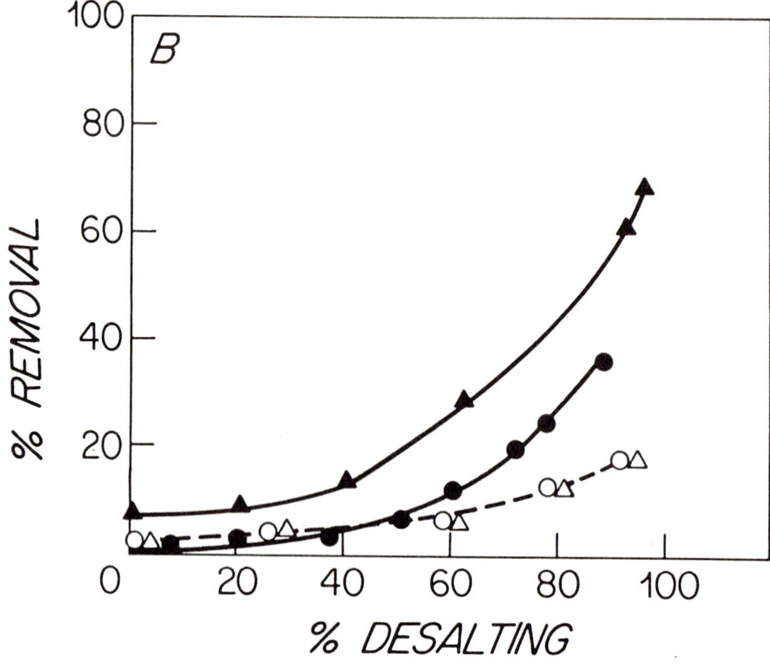

Figure 5: *Removal of exogenously added* [125]*I-labeled IgG and heat-aggregated IgG from CPD plasma as a function of salt removal by electrodialysis. The concentration of* [125]*I-labeled IgG (△, ○) was 30 μgm/ml, the* [125]*I-labeled heat-aggregated IgG (▲) was 30 μgm/ml, and 50 ml of plasma was used. An i/N=500 was used and 2.5% Na citrate was the electrode stream, pH 5.0 (▲, △) or 9.2 (●, ○), the concentrate stream was 0.25% NaCl, the temperature was maintained at 10°C, and 6 cell pairs were used. Electrodialysis was done in the Dial-a-Cell.*

from IgG. At concentrations of 50 mM there was 40–20% removal of IgM and variable recovery of IgA.

Plasma precipitated with 50 mM zinc diglycinate and then centrifuged at 2,000 × g was subjected to electrodialysis and these results are summarized in Table 2. By combining the zinc diglycinate precipitation with dialysis, it was possible to maintain high recovery of albumin (70%) while removing greater than 98% of the IgG and greater than 80% of the IgM

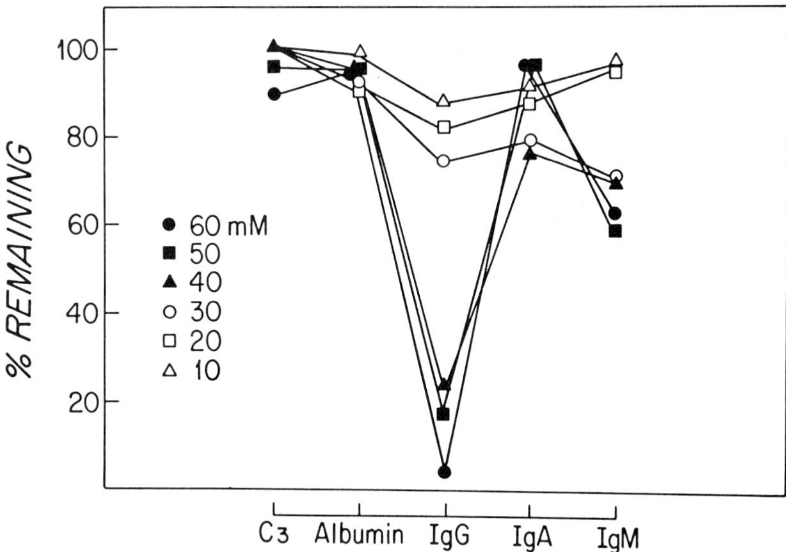

Figure 6: *Precipitation of selective plasma proteins from CPD plasma by varying concentrations of zinc diglycinate. Plasma (10 ml) was adjusted to various zinc diglycinate concentrations with a 0.5 M stock solution of zinc diglycinate. The pH was adjusted to pH 7.4 with 0.9 M NaHCO₃ and the precipitate removed immediately by centrifugation at 2,000 g for 10 minutes. The excess zinc was removed by dialysis and the samples were assayed immunochemically for the proteins. The % remaining was based on the initial values in untreated plasma and corrected for the volume change which occurred during precipitation and dialysis.*

(see column III, Table 2). There was variable but overall low recovery of the other measured plasma proteins when the two techniques were combined. The ability of zinc diglycinate to remove IgG/antibody and/or heat-aggregated IgG from plasma was further examined by adding exogenous TIG, or ^{125}I-IgG and heat-aggregated IgG to plasma prior to precipitation. All detectable antibody was removed by 50 mM zinc diglycinate (Table 3). The combination of zinc diglycinate with electrodialysis removed 88–93% of the exogenously added radiolabeled IgG (Table 4) and confirmed the results on precipitation of IgG from normal plasma by this reagent.

Table 2

Serum Protein Analysis of Proteins in Supernatants of CPD Plasma Following Precipitation with 50 mM Zinc Diglycinate Precipitation

Protein	*Electrode Stream 2.5% Na citrate, pH 5.5*					*Normal Values for Plasma Proteins**	
	I	*II*	*%Dec.*	*III*	*%Dec.*		
Albumin	3.2	2.9	9.4	2.2	31.3	3.4−4.5	g/dl
α_1-Antitrypsin	60	48	20	35	41.7	47−153	%nl
Haptoglobin	30	22	26.7	11	68.3	30−160	mg/dl
Transferrin	190	175	7.9	80	57.9	205−374	mg/dl
Orosomucoid	88	61	30.7	40	54.5	38−147	%nl
C4	60	56	66.7	42	30	25−175	%nl
C3	54	49	9.3	18	66.7	44−177	%nl
IgG	560	42	92.5	<10	>98	600−1500	mg/dl
IgA	125	119	4.8	40	68	60−290	mg/dl
IgM	54	21	61	<10	>82	50−200	mg/dl
Properdin factor B	25	20	20	10	60	12−56	mg/dl
β-Lipoprotein	62	38	38.7	18	71	50−135	mg/dl
Factor X	100	88[†]	12	95[†]	5		
Factor V	100	92[†]	8	98[†]	2		
Factor VIII	100	94[†]	6	98[†]	2		

*Ranges are two standard deviations from the mean of a pool of normal adult plasma (see *Methodology* section).

[†]Determined at a final concentration of 1% plasma. The initial value of the starting plasma was set as 100%

I = Initial Value

II = After zinc diglycinate precipitation

III = After electrodialysis of zinc diglycinate treated plasma.

Table 3

Percent Remaining of ^{125}I-IgG and ^{125}I-Heat-Aggregated IgG (ΔIgG) from Normal CPD Plasma

		Zn Diglycinate Precipitation	
	Electrodialysis	*Zn Diglycinate Alone*	*Zn Diglycinate Precipitation Followed by Electrodialysis*
^{125}I-IgG	82	27	12
^{125}I-IgG	33	13	7

^{125}I-IgG was added to 16 mg/dl (1.6×10^4 cpm/ml) and ^{125}I-IgG to 16 mg/dl (1.7×10^4 cpm/ml). Electrodialysis was performed with 2.5% Na citrate in pH 5.5 in the electrode stream. Precipitation was at a final concentration of 50 mM Zn diglycinate (see *Methodology* section).

Table 4
Removal of Exogenously Added Human Tetanus Toxoid Antibody
by Zinc Diglycinate Precipitation

Zinc Diglycinate (mM)	Hemagglutination Titer
0	32
10	32
20	32
30	32
40	32
50	<2
60	≤2
50	≤2*
50	≤2†

*No antibody added to plasma.
†Antibody added to plasma, but unsensitized red cells used.
After removal of the zinc diglycinate precipitate, antibody titers were determined on the supernatants by passive hemagglutination with glutaraldehyde-preserved sheep red blood cells sensitized with tetanus toxoid.

Effect of Electrodialysis and Precipitation with Zinc Diglycinate on C3 Activation

Because no significant changes in clotting activities were noted upon electrodialysis or zinc diglycinate precipitation, it was of interest to determine whether either procedure might activate complement, particularly since Cl, Clq, and immune complex removal was observed. For this experiment, fresh plasma collected into ACD anticoagulant was used, and samples subjected to electrodialysis with the electrode stream at pH 5.5 were compared to plasma precipitated with 50 mM zinc diglycinate. These samples were compared to separate plasma samples (1) recirculated through the ED cell, (2) treated with zymosan and heat-aggregated IgG, and (3) left standing at room temperature for the time required to do the electrodialysis/zinc diglycinate precipitation. The untreated sample and recirculated sample contained 108 and 105 mg% of C3 per dl. The zinc diglycinate removed a little over 50% of the C3 (49.5 mg/dl). Electrodialysis depleted 94% of the C3.

However, in all cases, the remaining C3 did not change in electrophoretic mobility nor was there a qualitative increase in C3 fragments upon examination of C3 in the plasma by crossed immunoelectrophoresis. In contrast, heat-aggregated IgG and zymosan-treated plasma (37°C for 1 hour) did not change C3 levels (105 and 107 mg/dl, respectively) but did cause considerable C3 activation (75% and 77%, respectively) as compared to plasma incubated alone at 37°C for 1 hour (see Fig. 7).

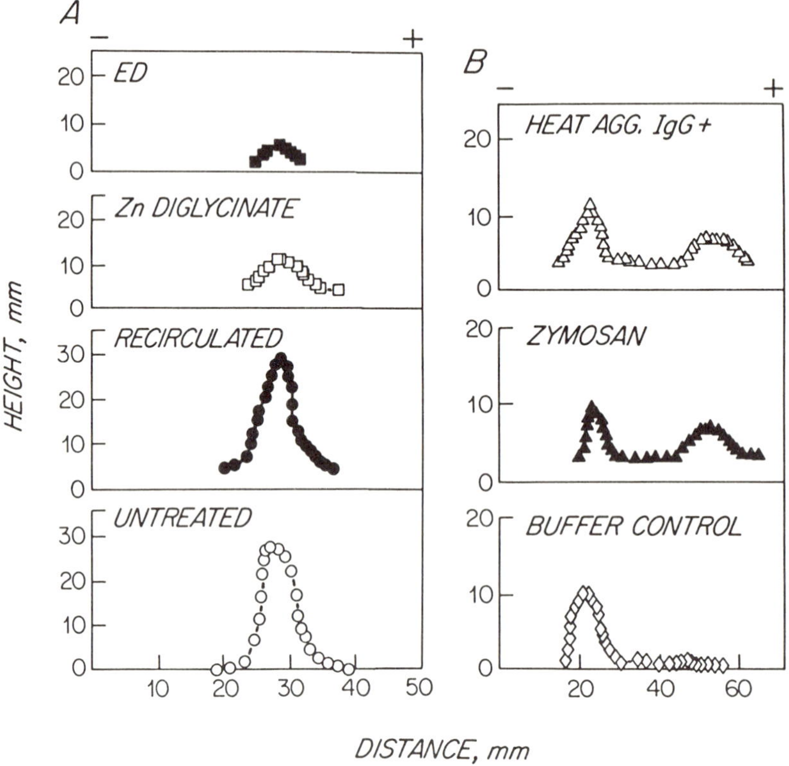

Figure 7: *Crossed immunoelectrophoresis of plasma treated with 50 mM zinc diglycinate. (**A**) (●) Untreated plasma; (○) plasma recirculated through the electrodialysis cell prior to deionization; (■) plasma electrodialyzed to 99% salt removal and centrifuged to remove precipitate; (□) plasma precipitated with 50 mM zinc diglycinate. (**B**) (△) Plasma incubated at 37°C for 1 hour with heat-aggregated IgG; (▲) plasma incubated at 37°C for 1 hour with zymosan; (◇) plasma incubated at 37°C for 1 hour with buffer.*

The Effect of Electrodialysis or Zinc Diglycinate Precipitation on Plasma Collected During Therapeutic Plasma Exchange

These studies were conducted with plasma obtained from patients undergoing therapeutic plasma exchange (TPE). In these experiments, only the initial 200–300 ml of plasma was examined, as in subsequent plasma bags proteins would be diluted with the replacement fluid. These plasmas were obtained in many cases some months after the therapy had been initiated, so no prospective analysis was possible. To test reproducibility of the technique, however, weekly samples were examined for some plasmas. Out of approximately 50 samples, plasmas collected from 12 patients with myasthenia gravis, 4 with Guillain-Barré syndrome, 2 with rheumatoid arthritis, 1 with systemic lupus erythematosus, 3 with gamma-globulinopathies, and 1 with acute transverse myelitis were examined.

The experimental design was as follows: The plasma was thawed at 15°C and maintained at this temperature. The plasma was treated by two methods. First, a 10 ml sample was precipitated with 50 mM zinc diglycinate after adjusting to pH 7.2 with 0.9 M $NaHCO_3$, pH 9.0. Second, a 100 ml sample was electrodialyzed under conditions where the pH of the plasma gradually dropped to 5 (see Fig. 1, electrode stream = 2.5% Na citrate pH 5.5, and Fig. 3). The following analyses were performed on all samples: total protein, albumin, IgG, IgM, IgA, and immune complexes. The data are summarized in Tables 5 and 6. While these plasmas have not been treated with the combined techniques, these separate data show that similar results are obtained on pathological plasma as with normal plasma. Namely, electrodialysis and selective chemical precipitation can lead to removal of immune complexes (where present), IgG, and a substantial amount of IgM, leaving the majority of the albumin and variable amounts of the other plasma proteins. Statistical analysis of the extent of removal by both methods is summarized in Table 7. IgM is the only protein that is not removed as well from the abnormal

Table 5
Effect of Electrodialysis on Selected Plasma Proteins in Plasma
Collected During Therapeutic Plasma Exchange

Patient	Disease	Date	Immune Complex[‡] HAG (μg/ml)			Albumin (g/dl)		
			Init.	Final	% Remain*	Init.	Final	% Remain*
GF	CB	09/21/81	195	95.50	49.0	1.34	1.30	93.1
		10/20/81	—[†]	—[†]	—[†]	—[†]	—[†]	—[†]
MR	HV	01/05/82	24.60	0.82	3.3	2.70	2.35	87
		10/20/81	35.70	1.53	4.3	2.41	1.81	76
PO	W	09/10/81	8.65	1.25	14.4	3.19	2.22	69.6
		01/19/82	6.59	0.94	14.3	3.51	2.69	76.6
		01/28/82	1.70	1.52	89.4	1.82	1.82	100
MC	RA	10/30/81	0.25	0.07	28.0	1.44	1.44	100
		11/30/81	0.64	0.36	56.3	—[†]	—[†]	—[†]
		02/03/82	—[†]	—[†]	—[†]	2.88	2.40	85
		12/30/81	1.08	0.38	35.2	1.53	1.42	93.1
		01/20/82	8.15	0.04	0.5	1.63	1.44	86.6
		02/16/82	1.47	1.34	91.2	3.59	2.81	78
CC	RA	07/20/81	0.32	0.11	34.6	1.36	1.34	98.5
		11/04/81	0.41	1.17	>100	1.29	1.38	107
		12/01/81	0.23	0.25	>100	1.51	1.16	77.3
		01/05/82	0.24	0.09	37	1.44	1.19	82.3
		01/26/82	0.31	0.56	>100	2.76	2.56	93
		02/02/82	—[†]	—[†]	—[†]	2.51	2.58	100
		02/16/82	—[†]	—[†]	—[†]	2.03	1.79	88

CO	MG	02/04/82	0.72	0.80	>100	4.33	3.96	91
		02/18/82	0.79	0.75	>100	1.99	2.09	105
MO	MG	07/14/81	0.79	0.80	>100	1.99	2.09	105
GV	MG	02/22/82	1.08	1.49	>100	3.01	2.78	92.4
OL	MG	07/09/81	0.40	0.53	>100	1.87	1.84	98.4
TR	MG	07/21/81	3.48	2.82	81	1.83	1.52	83.1
KI	MG	01/13/82	5.94	1.07	18	4.95	4.94	100
BR	MG	01/17/82	4.37	7.92	>100	—[†]	—[†]	—[†]
GR	MG	07/07/81	3.06	0.69	22.5	2.80	2.16	77.1
		02/02/82	6.49	2.45	37.5	1.98	2.07	104
		02/12/82	1.70	0.92	54.1	3.36	2.70	80
ST	GB	01/15/82	0.75	0.83	48.8	3.36	3.36	100
MU	GB	11/02/81	5.74	4.01	69.7	2.55	2.28	89.9
GN	GB	10/29/81	1.29	.87	67.6	2.19	1.66	75.8
CR	GB	10/20/81	38.90	1.85	0.93	3.24	3.03	93.8
DP	SLE	08/24/81	3.84	3.64	94.8	2.12	2.03	95.8
		09/30/81	5.40	2.80	51.9	2.59	2.47	95.4
		09/23/81	7.26	2.69	37.1	2.70	2.47	95.4
			0.72	0.32	44.7	2.49	—	—

[*]Final values were determined on samples from which the precipitate was removed by centrifugation (10 min at 5,000 × *g*) following electrodialysis.

[†]Omitted from analysis—see Table 7.

[‡]Normal values for immune complexes as determined at the CBR is 5.6 µg HAG/ml.

CB = cryoglobulinemia; HV = hyperviscosity; W = Waldenstrom macroglobulinemia; RA = rheumatoid arthritis; MG = myasthenia gravis; GB = Guillain-Barré syndrome; SLE = systemic lupus erythematosus.

Table 6
Effect of Zinc Diglycinate Precipitation on Selected Plasma Proteins During Therapeutic Plasma Exchange

Patient	Disease	Date	Total Protein (g/dl)			Albumin (g/dl)			IgG (mg/dl)			Patient	Disease	Date	IgM (mg/dl)			IC* HAG (µg/ml)	
			Init.	Final	%Remain.	Init.	Final	%Remain.	Init.	Final	%Remain.				Init.	Final	%Remain	Init.	Final
GF	CB	10/20/81	7.86	1.67	26.3	1.80	1.67	92.8	6160	80	1.3	GF	CB	10/20/81	14	10	71.4	195	0.40
MR	HV	01/05/82	5.60	3.40	60.7	2.70	2.33	86.3	≤20	≤20	—	MR	HV	01/05/82	2160	1520	70.4	24.60	35.50
PO	W	01/19/82	6.85	2.00	29.2	1.80	1.70	94.4	1600	≤20	< 1	PO	W	01/19/82	1280	192	15†	8.65	0.38
MC	RA	10/30/81	3.40	2.80	82.4	2.80	2.60	92.9	220	≤20	9.1	MC	RA	10/30/81	32	32	100	0.25	0.14
CL	RA	12/01/81	3.25	2.20	67.7	2.35	2.21	94.0	220	—	—	CL	RA	12/01/81	32	30	93.7	1.08	0.56
MO	MG	07/14/81	3.50	3.45	98.6	3.50	3.20	91.4	620	45	7.3	MO	MG	07/14/81	50	44	88	1.08	0.56
GV	MG	07/22/81	—	—	—	3.80	3.30	86.8	80	≤10	<12.5	GV	MG	07/22/81	24	24	100	0.40	0.31
OL	MG	07/09/81	2.60	2.48	95.8	2.30	2.10	91.3	700	50	7.1	OL	MG	07/09/81	50	40	80	3.48	0.60
TR	MG	07/21/81	5.05	2.47	48.9	2.60	2.47	95.0	260	≤20	< 7.7	TR	MG	07/21/81	72	50	69.4	5.94	1.00
KI	MG	01/13/82	4.15	3.14	75.7	3.20	3.14	98.2	740	120	16.2	KI	MG	01/13/82	58	56	90	4.37	4.80
GR	MG	07/07/81	2.50	2.40	96.0	2.50	2.80	>100	50	≤10	< 20*	GR	MG	07/07/81	148	128	86.5	1.85	1.85
BR	MG	11/17/81	4.80	2.20	45.8	1.70	1.64	96.5	330	≤20	< 6	BR	MG	11/17/81	34	34	100	3.06	0.47
ST	GB	03/12/82	2.34	1.50	64.1	1.40	1.20	85.7	200	≤10	< 5	ST	GB	03/12/82	12	10	83.3	5.74	0.50
MU	GB	11/02/81	2.30	—	—	1.10	0.80	72.7	130	≤10	< 8	MU	GB	11/02/81	≤10	≤10	—†	1.29	0.57
GN	GB	06/25/81	4.30	3.20	74.4	3.20	3.19	99.7	940	≤20	< 2	GN	GB	06/25/81	80	80	100	38.90	2.19
CR	GB	10/29/81	4.00	1.80	45.0	1.60	1.30	81.3	220	≤20	< 9.1	CR	GB	10/29/81	62	60	96.8	3.84	2.24
DP	SLE	09/23/81	4.50	2.26	50.2	2.30	2.26	98.3	940	≤20	< 2.4	DP	SLE	09/23/81	34	32	94.1	7.26	0.50

Final values were determined on samples from which the precipitate was removed by centrifugation (10 min at 5,000 $\times$ g) following zinc precipitation.
CB = cryoglobulenemia; HV = hyperviscosity; RA = rheumatoid arthritis; MG = myasthenia gravis; GB = Guillain-Barré syndrome;
SLE = systemic lupus erythematosus.
*Normal values for immune complex as determined at CBR was 5.63 µg HAG/ml.
†Omitted from analysis—see Table 7.

Table 7
Statistical Analysis of Albumin and Ig Recoveries on Plasmas
Collected During Therapeutic Plasma Exchange

Treatment	Disease	Total Protein*	n	Albumin*	$n^\dagger$	IgG*	$n^\dagger$	IgM*	$n^\dagger$
Electrodialysis	G	33.9 ± 9	6	79.9 ± 17.9	6	63.6 ± 33.5	6	25.5 ± 10.9	5
	MG	66.9 ± 14	11	93.6 ± 10.5	10	68.2 ± 27.9	11	52.2 ± 31	9
	GB	62.3 ± 11	4	89.9 ± 10.3	4	91.1 ± 18.6	4	62.7 ± 17.5	4
	RA & SLE	68.7 ± 13.4	14	91.7 ± 8.6	15	67.3 ± 20.8	15	30.5 ± 15.1	13
	All Data	61.5 ± 17.7	35	90.7 ± 10.0	35	69.6 ± 25.5	36	40.2 ± 24.3	31
Zinc Diglycinate	G	38.7 ± 19.1	3	91.2 ± 4.3	3	1.2	2	70.9	2
	MG	83 ± 21.2	7	94.2 ± 4.3	7	9.5 ± 4.0	6	87.7 ± 1.0	7
	GB	61.2 ± 14.9	3	84.5 ± 11.3	4	6.0 ± 3.1	4	93.4 ± 8.6	3
	RA & SLE	66.8 ± 16.1	3	95.1 ± 2.9	3	5.7	2	95.9 ± 3.5	3
	All Data	64.1 ± 23.3	15	91.6 ± 7.2	17	6.75 ± 4.3	14	88.2 ± 11.2	15

G = gammaglobulinopathies; MG = myastheria gravis; GB = Guillain-Barré syndrome; RA = rheumatoid arthritis; SLE = systemic lupus erythematosus.
*Mean ± SD.
†Number of determinations.

plasmas as from the normal. The data indicate that different diseases may behave differently but the number of measurements are too small to derive any firm conclusions on this point.

DISCUSSION

In this rather narrow study, we examined for two physicochemical protein fractionation methods certain parameters which could determine applicability of using the techniques to recover and use an individual's own albumin as the replacement fluid in therapeutic plasma exchange (TPE). Questions asked included (1) would the techniques be adaptable to the current plasmapheresis technology (e.g., centrifuge and membrane-filtration techniques for separation of plasma from the formed elements)?; (2) did the techniques remove sufficient quantities of IgG, IgM, and immune complexes while retaining sufficient levels of albumin to warrant further study of such plasma as possible replacement fluid in TPE?; (3) did the techniques qualitatively alter the nature of the retained plasma proteins such as the coagulation and complement components?; (4) would the techniques yield similar protein recoveries, at least on a percentage basis, with both abnormal and normal plasma?

Based on the results of this study, both of the techniques appear to be adaptable to current TPE therapy. The electrodialysis apparatus is modular in design. It thus can be expanded to accommodate large volumes of plasma, the flow rates and rate of deionization can be varied to meet any fixed value recognized by a given blood separation device, and it operates equally well from 0° to 37°C. Electrodialysis has the further advantage of not being dependent on concentration gradients to effect ion removal, and the degree of deionization can be exactly controlled. Although we have not tested whether electrodialysis can cause undesired side effects in vivo, the results of serum protein analysis indicate that, at least in normal plasma, there are no changes in functional activity of the retained factor X, factor VIII, and the C3 (Tables 1, 2 and Fig. 7). By such criteria, the usefulness of zinc diglycinate as a selective precipitating agent for IgG without altering retained coagulation proteins and C3 (Table 2 and Fig. 6) is also apparent. A potential problem with using zinc diglycinate is the removal of unprecipitated metal ion complex from the plasma. However, in the preparation of SPPS, by Method 12, ion exchange resins were used to remove barium salts and zinc diglycinate, and it would appear that the proven safety of Method 12 would mean that the technique could be adapted for TPE.

Neither electrodialysis nor zinc diglycinate precipitation alone will remove the IgG or IgM immune complexes while retaining all of the albumin in the plasma. Electrodialysis alone can lead to removal of immune complexes and 70% of the IgM. Zinc diglycinate removes IgG and immune complexes. With normal plasma, combining the techniques appears to give maximum separation of albumin and immunoglobulin/immune complexes (see Table 2), but both techniques lead to variable plasma protein losses which may be undesirable. In current TPE therapy, however, all the plasma proteins are discarded. Thus, while the techniques described in this study may not be as ideal, as, for example, an immunoadsorbent, in terms of not disturbing levels of the normal plasma proteins in an abnormal plasma, there is at least a measure of retention of the overall protein content of the plasma (see Tables 5 and 6). Furthermore, the concentrations of plasma proteins retained and removed are quite reproducible. The data presented in Tables 1 and 2 on normal CPD plasma are representative of over 50 experiments. An analysis of albumin recovery and IgG and IgM removal on the abnormal plasma (see Table 7) further demonstrates the reproducibility of both electrodialysis and zinc diglycinate precipitation in spite of the heterogeneous nature of these plasmas.

The rationale for initiating this study was based on the proven ability of physicochemical techniques to reproducibly remove immunoglobulins and immune complexes from plasma. This study has focused on two such methods—deionization of undiluted plasma (e.g., isoelectric precipitation) and selective precipitation of IgG with a metal ion complex. There may be other techniques such as dialysis, ion exchange resins, and molecular sieving that could be effective as well. Our choice of these methods has been based on our ability to control the process of fractionation, the potential compatibility of the methods with blood separation devices used in plasmapheresis, and the reproducibility of the method when applied to plasmas derived from individuals undergoing therapeutic plasmapheresis. The present results indicate that separations can be obtained by these techniques and that these results warrant further in vivo studies to determine whether the methods will be efficacious for those patients for whom therapeutic plasma exchange has proven to be a useful therapy.

SUMMARY

Two physicochemical plasma fractionation techniques—isoelectric precipitation by rapid deionization by electrodialysis and precipitation with the metal ion complex, zinc diglycinate—have been examined as techniques that can remove immunoglobulins and exogenously added immune complexes/heat-aggregated IgG from normal plasma. The results of studies on normal plasma have been compared to results of depletion of immunoglobulins and endogenous immune complexes from plasma obtained from individuals undergoing therapeutic plasma exchange. Isoelectric precipitation can remove immune complexes/heat-aggregated IgG and up to 70% of the IgM from plasma. Zinc diglycinate removes greater than 90% of both the IgG and the immune complexes. Both techniques displace very little of the endogenous albumin (90% to 80% retained) and affect neither the remaining coagulation proteins nor C3. Combination of the two physicochemical methods with conventional methodologies for plasmapheresis should make it feasible to use a patient's own immunoglobulin/immune complex depleted plasma as the replacement fluid in therapeutic plasma exchange.

REFERENCES

1. Nemo GJ, Taswell H: *Proceedings of the Workshop on Therapeutic Plasmapheresis and Cytapheresis.* NIH Publication No. 82-1665. Washington DC, Government Printing Office, 1981.
2. Bansal SC, Bansal BR, Rhoads JE Jr, et al: Ex vivo removal of mammalian immunoglobulin G: methods and immunological alterations. *Int J Art Organs* 1980; 1:94.
3. Terman DS: Extracorporeal immune adsorbents for therapy of autoimmune and neoplastic disease. **In** *Proceedings of the Workshop on Therapeutic Plasmapheresis and Cytapheresis.* Edited by GJ Nemo, H Taswell. NIH Publication No. 82-1665. Washington DC, Government Printing Office, 1981, p 301.
4. Cantor CR, Schimmel PR: *Biophysical Chemistry. Part I. The Conformation of Biological Macromolecules.* San Francisco, CA, W Freeman, 1980, p 279.
5. Richards FM, Richmond T: Solvents, interfaces and protein structure. *Ciba Found Symp* 1978; 60:23.
6. Haschemeyer RH, Haschemeyer AEV: *Proteins: A Guide to Study in Physical and Chemical Methods.* New York, John Wiley & Sons, 1973, p 368.
7. Schultze HE, Heremans JH: *Molecular Biology of Human Proteins With Special Reference to Plasma Proteins.* New York, Elsevier Biomedical Press, 1966, p 236.
8. Cohn EJ, Edsall JT: *Proteins, Amino Acids and Peptides: As Ions and Dipolar Ions.* New York, Reinhold Publishing Corporation, 1943.
9. Von Hippel PH: Neutral salt effects on the conformational stability of biological molecules. **In** *Protein-Ligand Interactions.* Edited by H Sund, G Blauer. New York, Walter De Gruyter, 1975, p 452.
10. Ahlgren PM: Electrodialysis through ion exchange membranes. *Ion Exch Poll Control* 1979; 1:145.
11. Jain SM: Electrodialysis and its applications. *Am Lab* October 1979.
12. Juda W, McRae WA: Coherent ion-exchange gels and membranes. *J Am Chem Soc* 1950; 72:1044 (letter to the editor).
13. Parsi EJ: Large electrodialysis stack development. *Desalination* 1976; 19:139.
14. Cohn EJ: Interactions of proteins with each other and with heavy metals. **In** *Blood Cells and Plasma Proteins: Their State in Nature.* Edited by JL Tullis. New York, Academic Press, 1953, p 29.

15. Isliker HC, Antoniades HN: Fractionation of antibodies from human γ-globulins with zinc salts. *J Am Chem Soc* 1955; 77:5863.

16. Surgenor DM: Recent advances in the preparation of stable plasma derivatives. *Q Rev Med* 1952; 9:145.

17. Surgenor DM, Pennell RB, Alameri E, et al: Preparation and properties of serum and plasma proteins, XXXV: A system of protein fractionation using zinc complexes. *Vox Sang* 1960, 5:272.

18. Pennell RB: Fractionation and isolation of purified components by precipitation methods. **In** *The Plasma Proteins: Isolation, Characterization, and Function.* Vol 1. Edited by FW Putnam. New York, Academic Press, 1960, p 9.

19. Bing DH, Andrews JM, Morris KM, et al: Purification of subcomponents Clq, Cl(-)r and Cl (-)s of the first component of complement from Cohn Fraction I by affinity chromatography. *Prep Biochem* 1980; 10:269.

20. Wehler C, Andrews JM, Bing DH: The use of solid phase Clq and horseradish peroxidase-conjugated goat anti-IgG for the detection of immune complexes in human serum. *Mol Immunol* 1981; 18:157.

21. Bourke BE, Moss IK, Mumford P, et al: The complement fixing ability of putative circulating immune complexes in rheumatoid arthritis and its relationship to extra-articular disease. *Clin Exp Immunol* 1982; 48:726.

22. Bing DH, Weyand JGM, Stavitsky AB: Hemagglutination with aldehyde-fixed erythrocytes for assay of antigens and antibodies. *Proc Soc Exp Biol Med* 1967; 124:1166.

23. McConahey PJ, Dixon FJ: A method of trace iodination of proteins for immunologic studies. *Int Arch Allergy Appl Immunol* 1966; 29:185.

24. Ritchie RF: Automated immunoprecipitation analysis of serum proteins. **In** *The Plasma Proteins.* Vol 2. Second edition. Edited by FW Putnam. New York, Academic Press, 1975, p 376.

25. Ritchie RF: Specific proteins. **In** *Clinical Diagnosis and Management by Laboratory Methods.* Vol 2. Sixteenth edition. Edited by JB Henry. Philadelphia, WB Saunders Company, 1979, p 228.

26. Laki K: The polymerization of proteins: The action of thrombin on fibrinogen. *Arch Biochem Biophys* 1951; 32:317.

27. Hardisty RM, Macpherson JC: A one-stage factor VIII (antihaemophilic globulin) assay and its use on venous and capillary plasma. *Thromb Diath Haemorrh* 1962; 7:215.

28. Rapp HJ, Borsos T: *Molecular Basis of Complement Action.* New York, Appleton-Century-Crofts, 1970, p 75.

29. Mancini G, Carbonara AO, Heremans JF: Immunochemical quantitation of antigens by single radial immunodiffusion. *Immunochemistry* 1965; 2:235.

30. Laurell C-B: Electroimmuno assay. *Scand J Clin Lab Invest* 1972; 29 (Suppl 124):21.

31. The PROPHET system. *Fed Proc* 1973; 32:1744.

32. Sandor G: *Serum Proteins in Health and Disease.* (Edited and translated from French by E Kawerau) London, Chapman & Hall, 1966, p 105.

CHAPTER 8

Selective Removal of Macromolecules by Cryofiltration

Takashi Horiuchi, M.S.
Paul S. Malchesky, M.S.
James W. Smith, M.D., Ph.D.
Yukihiko Nosé, M.D., Ph.D.

INTRODUCTION

The success of hemodialysis in the treatment of acute and chronic renal failure has promoted the development of extracorporeal technology for the therapy of various other diseases. Hemodialysis has been established as a safe and effective methodology based upon several decades of application, and its use has spurred many other technologic developments in the medical field. Major improvements have occurred in the past decade in adsorbents for use in the treatment of drug intoxication and hepatic support, in ultrafiltration membranes for the removal of middle molecular weight substances, and in microporous membranes for the separation of plasma from whole blood. Diseases believed to be mediated by macromolecular weight substances such as circulating immune complexes, antibodies, protein bound toxins, and other abnormally occurring molecules that accumulate in disease states have been treated by blood or plasma exchange since 1960.[1-5]

Cell centrifugation techniques have been utilized mainly in acute application for therapeutic exchange of plasma despite limitations of blood cell and plasma component losses, the incomplete separation of plasma from blood cellular elements, and the requirement for plasma products and complications associated with their infusion, e.g., allergic reactions or infection.[6,7] Based upon early success in the treatment of various diseases by plasma exchange,[8-16] its use in many diseases, and the use of the therapy for chronic treatment, the need for plasma products will increase. This will stimulate the development of therapies that do not require plasma products.[17-19] From the practical point of view, minimizing the need for plasma products can allow plasma treatments to be carried out more frequently, more extensively (processing larger volumes), and at lower costs. Because essential plasma components, as well as pathological ones, are removed during plasma exchange, techniques that are designed to remove only the pathological components are highly desirable. Techniques that have been investigated include sorption, physicochemical treatments, and filtration. The objectives of this paper are to review the various plasma treatment schemes with emphasis on the technique of cold filtration (cryofiltration) which was introduced by Nosé and Malchesky.[20-23]

PLASMA AVAILABILITY

The major deterrent to plasma exchange, whether performed by centrifuges or membranes, is the requirement for plasma products. In 1981, just over 6 million liters of plasma were collected in the United States to supply an international marketplace[24] as shown in Table 1. In plasma exchange, the plasma separated from whole blood is discarded and requires the infusion of a replacement fluid ranging from a colloidal physiological solution to fresh donor plasma (Table 2). If plasma exchange is successful in the treatment of the wide variety of disease states being investigated, competition for the available supply of plasma products would place serious

Table 1
Sources and Use of Plasma in the United States (Million Liters)[24]

Sources of Plasma	1975	1977	1979	1981	1983*	1985*	1987*
Totals from whole blood for transfusion as whole plasma; for fractionation	1.1	1.5	1.8	2.1	2.4	2.7	3.0
From plasmapheresis	2.0	2.6	3.4	4.3	5.5	7.0	9.0
Imported from third world countries	0.3	0.2	-	-	-	-	-
Total plasma collected	3.4	4.3	5.2	6.4	7.9	9.7	11.9
Uses of Plasma							
Total plasma for fractionation from whole blood, from plasmapheresis, and imports	3.0	3.6	4.3	5.4	6.7	8.3	10.2
For transfusion	0.4	0.5	0.5	05	0.5	0.5	0.5
Exported to Europe and Japan	-	0.2	0.4	0.5	0.7	0.9	1.2
Total Usage	3.4	4.3	5.2	6.4	7.9	9.7	11.9

*Projected

Table 2
Comparison of Plasma Replacement Fluids

Source	Advantages	Disadvantages
Albumin	Depletion of inflammatory mediators	High cost Depletion of other plasma constituents
Plasma protein fraction	Ease and convenience of use	High cost Reduction in coagulation factors
Fresh frozen plasma	Maintains normal plasma factors	Hepatitis Allergic reactions
Crystalloids	Low cost	Protein depletion

constraints on this form of treatment. Measures to encourage greater donor participation could perhaps help meet the growing need; however, in light of the increased demand (nearly 15% per year) for plasma fraction alone,[25] the needed supply for chronic plasma exchange programs cannot be guaranteed. Market surveys indicate that 500,000 to 750,000 patients could be treated annually. Total treatment numbers may be 10 times the patient population. Thus, while plasma exchange is very promising as a treatment modality for metabolic and immunologic disorders, this technique is not practical or cost effective for widespread applications. Aside from availability, another factor limiting application of plasma exchange is the removal of large amounts of macromolecules from the patients and the infusion of large amounts of foreign macromolecules. Since many of these patients' immunologic systems are abnormal, a negative impact on the pathophysiological state of the patient is easily conceivable.

PLASMA TREATMENT SYSTEMS

The preferred method of plasma treatment would be to remove only those factors associated with the disease state, thereby preventing the loss of essential substances in the plasma and eliminating the need for plasma product infusion.

Sorption

Development and use of sorbents in direct hemoperfusion for drug overdose, uremia, and liver insufficiency have been carried out over the past two decades.[26] Problems of biocompatibility, the potential problems of sorbent carryover into the vascular system, and the limited and nonspecific sorption characteristics of the sorbents limit their use. To overcome the problems of biocompatibility and sorbent carryover, the on-line separation of plasma and its subsequent treatment by sorbents has been used by our group since

1974.[27-43] This work has been oriented toward the treatment of acute and chronic liver failure. For the treatment of autoimmune disorders, immunosorbents, which are specific for the removal of immunoproteins, are being developed and clinically applied for use in on-line plasma treatment systems or direct hemoperfusion[44-58] as summarized in Table 3. In studies with immunosorbents to date, great concern has been expressed over the release of biologically active materials. In the design of specific adsorbents in which biological materials are immobilized on solid supports, it is important that carryover of the biological material does not occur. While this technology may be very promising for selected disease states, in most immunological diseases, multiple abnormalities exist and multiple sorbents may be required. In addition, the etiology of the disease is unknown in most instances, making the selection of the proper sorbent system difficult. Only limited clinical trials have been carried out with immunosorbents, and many of those were in the investigators' laboratories.

Physicochemical Methods

Various physicochemical methods for plasma treatment (employed primarily for analytic purposes) include deionization or electrodialysis [59] to remove salts and precipitate proteins, with subsequent dialysis to restore the electrolyte content. Other techniques used include the addition of agents to promote the precipitation of specific salts—e.g., polyethylene glycol for the complexing of immunocomplexes[60] and sulfated polysaccharides for the complexing of lipoproteins.[61] Such techniques, however, have not yet become clinically feasible.

Membrane Plasma Filtration

A simpler treatment scheme and one more universal in application would consist of the nonspecific removal of macromolecules in plasma, including the removal of those solutes

Table 3
Immunoadsorption Systems Described in the Literature[17]

Immunoadsorbent	Substance Removed	Application	Author
Protein A	IgG IgG immune complexes	In vitro	Rawer et al.[50]
Synthetic blood group B antigenic determinants	Anti-B antibodies	In vitro	Chang[48]
Glomerular basement membrane antigen	Anti-GBM antibodies	Dogs	Terman et al.[56]
Bovine serum albumin	Anti-BSA antibodies	Dogs	Terman[54]
Protein A	IgG	Dogs	Terman et al.[57]
Heparin agarose	LDL	Dogs	Burgstaler et al.[47]
Anti-LDL antibodies	LDL	Pigs	Stoffel et al.[53]
Glutaraldehyde treated human erythrocytes	Complement components	Sheep	Schmer et al.[51]
Heparin agarose	LDL	Sheep	Schmer et al.[52]
Staphylococcus aureus Cowan I	IgG	Human	Bansal et al.[44]
DNA	Anti-DNA antibodies	Human	Terman et al.[55]
Staphylococcus aureus Cowan I	IgG	Human	Ray et al.[49]
Blood group A and B antigen	ABO antibodies	Human	Bensinger et al.[45]
Anti-LDL antibodies	LDL	Human	Borberg et al.[46]
Protein A	IgG	Human	Terman[58]

followed as markers of the disease state. Review of the disease states treated by plasma exchange reveals that many of the marker solutes are larger than albumin (generally greater than 100,000 daltons) suggesting filtration as a physical separation technique for their removal (Fig. 1).[62] Membranes hav-

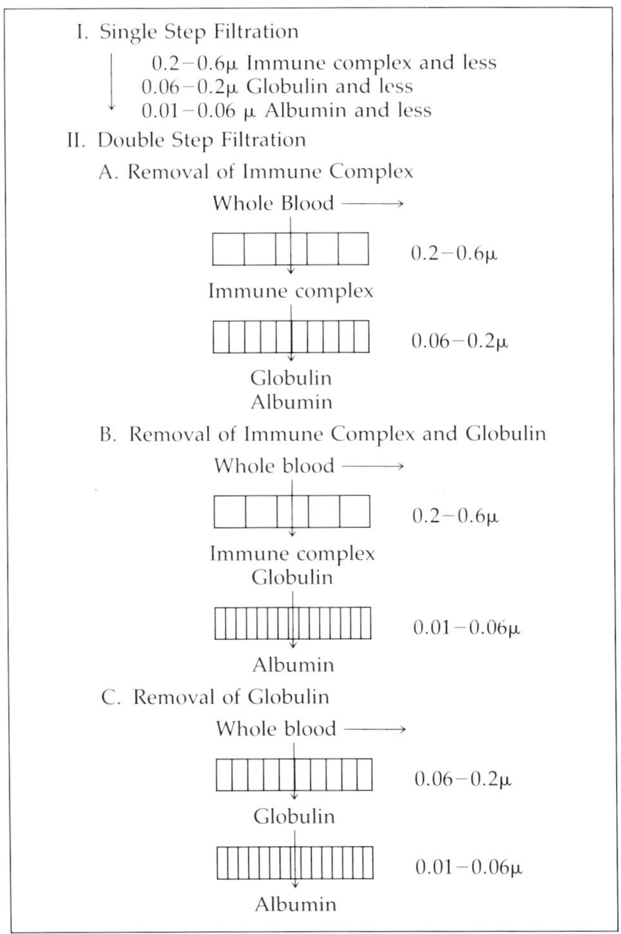

Figure 1: *Cascade membrane technology for selective removal of macromolecules. This scheme requires the selection of membranes of various porosities suitable for the selective removal of macromolecules from blood. (From Nosé Y, et al: Hybrid artificial organs: Are they really necessary? Artif Organs 1980, 4:285. By permission of the International Society for Artificial Organs.)*

ing a molecular cut off about 100,000 daltons and over can be selected, allowing the passage of albumin while retaining the macromolecular weight marker solutes (Table 4).

Selective removal of macromolecules by cascade membrane technology is theoretically possible. Efforts were begun in 1975 by our group for the development of cascade membrane systems.[23]

Evaluations of different membranes over a range of nominal pore sizes show that in practice, 100% passage of albumin and 100% rejection of macromolecules such as immune complexes and immunoglobulins are not achievable. Selectivity, particularly between the smaller globulin fractions and albumin, is difficult to achieve. In comparing the size differences of the various plasma proteins, it is noted that while molecular weights may vary widely, size and shape factors may make them more similar. For example, while albumin and IgG differ

---

Table 4
Ten Major Objectives of Secondary Filter (Macromolecule Filter)
for Albumin Regeneration[85]

1. On-line processing of plasma for albumin filtration (molecular weight less than 100,000) up to 3 liters at a plasma flow rate of 30 ml/min.

2. Over 80% of albumin should be filtered in the filtrate.

3. High extraction of pathologic macromolecules.

4. Should be able to filter plasma in which macromolecule (e.g., immune complex) concentration may be significantly higher than normal concentration.

5. Should be leak-free at inlet pressure of over 600 mm Hg.

6. Results should be reproducible.

7. Negative denaturation of plasma constituents.

8. Should be able to perform at near freezing environment.

9. Stable operating conditions with wide range of macromolecule levels in plasma.

10. Possible mass production at reasonable cost.

(Modified from Nosé Y, Malchesky PS: Technical Aspects of Membrane Plasma Treatment. *Artif Organs* 1981; Suppl 5:86. By permissin of the International Society for Artificial Organs.)

by a factor of 2 in molecular weight, shape factors and molecular orientation of the molecule at the filtering surface produce similar sieving properties in solutions containing a mixture of each. While the double membrane and cascade filtration schemes [63-67] can be designed to substantially pass albumin and reject macromolecules, the efficiency of the system is highly dependent upon the selection of the membrane, the choice of macromolecules to be removed, and the operating conditions. [68,69]

Cryofiltration System

It is well known that various disease states include cold-precipitable serum substances. While cryoprecipitates are seen mainly in a wide variety of immunological diseases, they can also be found in infectious diseases, neoplastic disorders, or in an essential form. The presence of cryoglobulins may represent a physiological part of the body's immune response, [70] which can evolve into a pathologic state depending upon the duration of stimulation and quantity and composition of the cryoglobulins synthesized. [71] Antigen as well as antibody activity has been demonstrated in cryoprecipitates. The presence of hepatitis B virus particles in cryoglobulins of patients with mixed cryoglobulinemia has been reported. [72] Antibodies to DNA in cases of lupus erythematosus, and rheumatoid factor in patients with rheumatoid arthritis have been found in the cryoprecipitates. [73,74] Antinuclear activity is neutralized after cryoprecipitation in the serum of patients with infectious mononucleosis. The presence of immune complexes and other immunological agents in cryoprecipitates suggests that the clinical response after cryoglobulin removal, as in plasma exchange, is primarily due to the decrease of immune complexes and other immunologic agents. Because of the association of cryoprecipitable solutes with a variety of disease states, off-line cryoglobulinpheresis[75] and cryopheresis[76,77] have been employed to remove these cryoproteins (Table 5). Plasma separated from whole blood by cell

Table 5
Comparison of Procedures in Cryotherapy
Cryofiltration: Cryoglobulinpheresis: Cryopheresis

Procedure (Disease state treated)	Anticoagulant	Plasma Separation	Cooling Time	Macromolecule Separation	System	Removal
Cryofiltration[21] (rheumatoid arthritis, cryoglobulinemia, SLE)	Heparin	Membrane plasma separator	10–12 minutes	Membrane plasma filter at 4°C	On-line	Cryogel
Cryoglobulinpheresis[75] (cryoglobulinemia)	ACD	Cell centrifugation	72–120 hours	Cell centrifugation 2,500 x-g for 15 min at 4°C	Off-line	Cryoprotein
Cryopheresis[76] (cryoglobulinemia)	ACD	Cell centrifugation	15–30 minutes	Cell centrifugation 3,000 x-g for 15 min at 0°C	Off-line	Cryoprotein

centrifuge is collected into sterilized bags and then cooled to 0−4°C. In cryoglobulinpheresis, after 72−130 hours of cooling, formed cryoprecipitates are separated by cell centrifugation and discarded. In off-line cryopheresis, cooling at 0°C for 15−30 minutes is employed prior to separation by centrifugation.

The technique of cooling the plasma prior to filtration was utilized to overcome the disadvantages of off-line systems for the removal of cryoprecipitable substances.[21] This technique of on-line plasma cooling and its filtration is referred to as cryofiltration. Cryofiltration does not remove cryoprotein alone; this process causes the formation and removal of cryogel, which is formed quickly in the chilled plasma. Cryogel includes not only cryoproteins but also various types of macromolecules. Decreases of temperature promote the gelation or aggregation of solutes, thereby increasing their overall molecular sizes and thus reducing the strict requirements of porosity for the filtering membrane and operating conditions. In in vitro studies with cryofiltration using plasma from patients with rheumatoid arthritis and rapidly progressive glomerulonephritis, effective removal of cryoprecipitable proteins and macromolecules with the high passage of albumin (Fig. 2) was shown. Figure 3 outlines schematically the circuit employed for cryofiltration. From the patient's point of view, the extracorporeal technique is similar to hemodialysis or plasma exchange in that anticoagulant and blood access is required. Blood flow rates typically employed are relatively low (50 ml/min to 200 ml/min). Blood access may be by cannulation of the superficial veins or arteriovenous fistula. Heparin is used as the anticoagulant in doses three to four times those typically given for hemodialysis. Important to the performance and efficiency of cryofiltration are the design and operation of the plasma separator, the macromolecule filter (cryofilter), and temperature control.

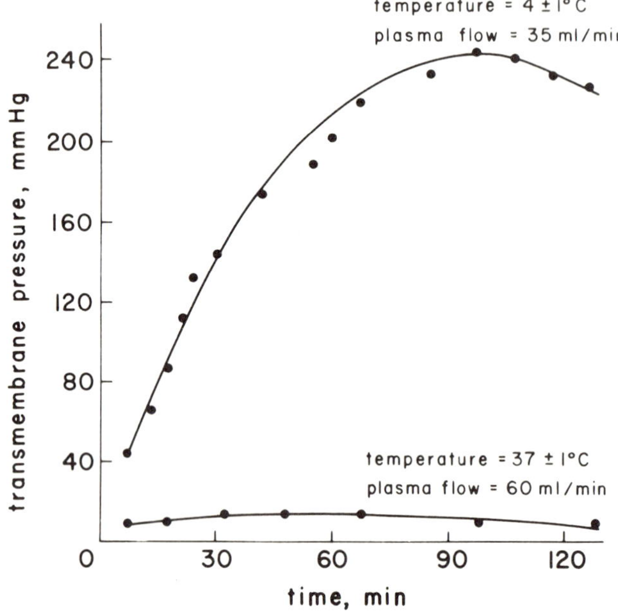

Figure 2: *In vitro cryofiltration test using plasma from a patient with rapidly progressive glomerulonephritis, indicating pressure change with time and temperature.*

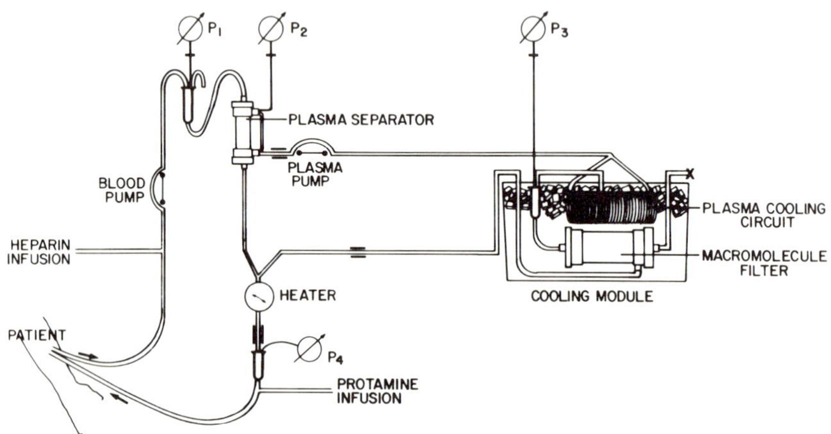

PLASMA SEPARATION FROM WHOLE BLOOD

The separation of blood cells from plasma is done routinely by centrifugal techniques.[78,79] For source plasma collection, the manual collection of blood with off-line separation is most often employed since the collection process cost is low and time of collection is not of critical importance. For therapeutic applications, on-line plasma separation is preferred. Available centrifugal systems are very effective for the separation of plasma from whole blood. However, costs for such on-line equipment are relatively high, the equipment is relatively complex, requiring the control of multiple parameters on a continuous basis by trained personnel, and the separation between the plasma and blood cells is not complete, thus creating disincentives to the widespread use of such technology. Appreciable carryover of blood cells, particularly platelets, occurs in the plasma.[7] In applications with on-line plasma treatment, these cells can interfere with the on-line processing, and their removal prior to the plasma treatment is required. To overcome these disadvantages of the centrifugal techniques, membrane technology was developed—initially plate type design[34,38] and later, hollow fiber design.[41] While membrane modules for plasma separation are just now becoming available in the United States for widespread clinical usage, such modules have been under clinical investigation since the late 1970s.[30] In general, the membranes used for this purpose have a nominal pore size of 0.1 to 0.5 μm. Some of the membranes have pores of uniform size while others contain pores of varying size randomly distributed as noted from surface structure studies.[80] Both hydrophilic and hydrophobic type membranes are being used. Hydrophobic membranes

---

Figure 3: *Scheme of on-line membrane plasmapheresis with cold filtration of plasma for the removal of macromolecules. (From Malchesky et al: On-line separation of macromolecules by membrane filtration with cryogelation. In* Plasma Exchange. *Edited by HB Sieberth. Stuttgart, FK Schattauer Verlag, 1980, p 133. By permission of publisher.)*

require pretreatment to wet the pores. Currently, the cellulose acetate hollow fiber membranes (Asahi Medical Co., Tokyo, Japan) are the most widely utilized. Other membrane materials include polyvinyl alcohol, polymethylmethacrylate, polypropylene, polyethylene, and polyvinyl chloride (Table 6).[81] Modules may be of flat plate or hollow fiber design. For flat plate design, most often blood channel controllers or adjusters are required. Hollow fibers utilized have diameters of 250 to 400 μm and a wall thickness of 50 to 200 μm.[2]

Blood is a mixture of various types of cellular components ranging from 1 to 15 μm and of plasma. Plasma contains various solutes of sizes ranging from macromolecules of a few million daltons in molecular mass to electrolytes of less than 50 daltons. Therefore, the separation of plasma from the cellular components requires special consideration in the operation of these devices as compared to hemodialyzers or ultrafilters (Fig. 4). In addition to consideration of the operating conditions, the chemical structure and microstructure of the filter media and the geometry and design of the separator are important to performance. Membrane modules vary in surface area from about 0.1 to 0.8 m^2. Membrane plasma separation is a relatively simple process; however, it requires special technical considerations. Technical aspects for the proper orientation of the plasma filter have been previously described in detail.[82] Adequate plasma fluxes in the absence of cellular carryover can be achieved at relatively low transmembrane pressures (generally less than 50 mm Hg). Equipment requirements are minimal and the operation is much like that for other extracorporeal treatment technologies such as hemodialysis, hemofiltration, and hemoperfusion. For blood flows of 50 to 200 ml/min (flows clinically employed), plasma flows of 15 to 70 ml/min are achieved. The passage (sieving) of macromolecular weight plasma solutes is over 90%.[83–85] The separation of plasma without cellular carryover has provided a further impetus for the development of on-line plasma treatment schemes, which minimize requirements for plasma products. Studies carried out with membrane systems

Table 6
Membrane Plasmapheresis Modules[81]

Trademark	Manufacturer	Material	Inside Diameter (μm)	Wall Thickness (μm)	Effective Length (mm)	Effective Surface Area (m²)	Priming Volume (ml)	Remarks
Asahi Hi-05	Asahi Med. (Japan)	Cellulose-diacetate	330	75	157	0.50	65	Hollow fiber, commercially available
Kuraray SA	Kuraray (Japan)	Polyvinyl alcohol	330	125	290	0.60	60	Hollow fiber, in clinical evaluation
Plasmax PS-05	Toray (Japan)	Poly methacrylate	370	85	175	0.50	70	Hollow fiber, in clinical evaluation
MPS	Mitsubishi Rayon Co. Ltd. (Japan)	Polyethylene	270	60	175	0.65	60	Hollow fiber, in clinical evaluation
Plasmaflux PS-500	Fresenius (Germany)	Polypropylene	330	140	200	0.50	55	Hollow fiber, commercially available in Europe
PS-510K	Takeda Ltd. (Japan)	Polypropylene	330	140	155	0.20	20	Hollow fiber, experimental unit
Pile type	Terumo (Japan)	Cellulose-diacetate	80*	170	45	0.40	80	Flat-type in preclinical evaluation

*Calculated value

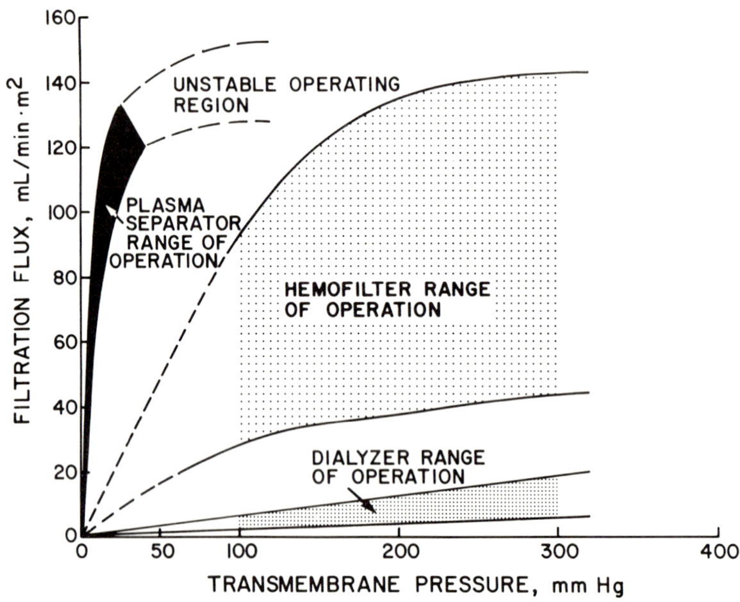

Figure 4: *Differences of optimum operating range between plasma separators, hemodialyzers, and hemofilters.*

indicate that they can be cost competitive with centrifugal methods, especially when used in conjunction with on-line plasma treatment schemes (Table 7).

Cryofiltration Process and Definition[86,87]

Plasma separated by the membrane plasma separator is continously pumped into the heat exchanger to lower the plasma temperature. In initial studies ice water was employed for cooling, while recent systems incorporate a thermo-electric cooler for plasma cooling (Cryomax, Parker Bio-medical, Irvine, CA) as shown in Figure 5. The cooled plasma is then filtered by the cryofilter (macromolecule filter for cryo-filtration). The Asahi Plasmaflo (Asahi Medical Co., Tokyo, Japan) originally investigated as a plasma separator, is used as

Table 7
Ten Major Objectives of On-Line
Membrane Plasma Processing Systems[85]

1. Ease of assembly for disposable units.

2. Total priming of circuitry less than 700 ml.

3. Effective cooling and warming of plasma.

4. Ease of operation.

5. Transmembrane pressure monitoring of primary filter for effective operation.

6. Incorporation of safety monitoring.

7. Possible incorporation of sorbent cartridge(s).

8. Ability to accommodate different types of various designs.

9. Provision for quick replacement of secondary filter during operation.

10. Mobile unit.

(Modified from Nosé Y, Malchesky PS: Technical Aspects of Membrane Plasma Treatment. *Artif Organs* 1981; Suppl 5:86. By permission of the International Society for Artificial Organs.)

the cryofilter (Table 8). The filtered plasma is united with the mainstream blood flow, and then warmed to physiological temperature before being returned to the patient. Priming volume of the plasma circuit is about 300 ml and that of the total system, about 450 ml. Plasma flow rates are about 30 ml/min at a blood flow of 100 ml/min and maintained constant until the transmembrane pressure across the cryofilter reaches 300 mm Hg. During clinical treatment, two cryofilters may be used to process 3 to 5 liters of plasma.

In cryofiltration, cryogel is formed and removed by the cryofilter.[88] The quantity of cryogel formed is higher at near freezing temperatures but cryogel formation may occur at ambient temperatures. Double filtration [63] also removes some cryogel because room temperature operation generally produced 25–30°C at the second filter level. Cryogel as generated in cryofiltration is distinct from cryoprecipitates or cryoprotein as isolated by classical laboratory tests or those substances

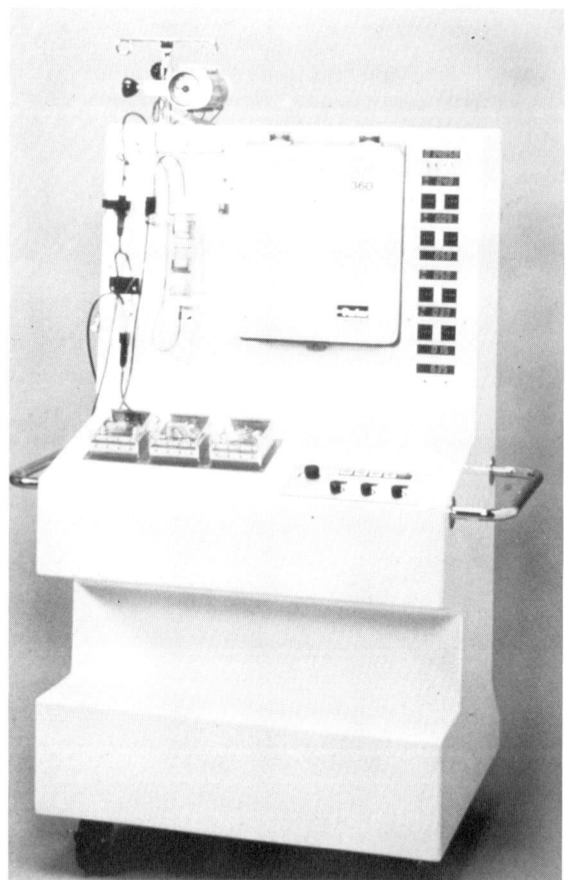

Figure 5: *Cryomax, an on-line system for plasma separation with cryofiltration, developed from the system devised at the Cleveland Clinic.*

removed by off-line techniques such as cryopheresis and cryoglobulinpheresis. Cryogel is formed from heparinized plasma at reduced temperatures in relatively short times (10–20 minutes) and is removed by membrane filtration. Several distinctions can be made between cryogel and cryoprecipitate, as shown by laboratory techniques (Table 9). Cryogel is formed from plasma, not serum, and is formed from a continuous flow stream. Membrane filtration removes the cryogel from

Table 8
List of Cryofilters Used Clinically

Unit Type	Manufacturer	Material	Surface Area (m^2)	Nominal Pore Size µm	Tp	Sieving Properties[*] Alb	Fib	IgG	IgM
Asahi Plasmaflo	Asahi Med. Co.	Cellulose Acetate	0.65	0.1−0.2	0.86	0.89	0.65	0.82	0.66
Kuraray MN	Kuraray Co.	Polyvinyl Alcohol	0.43	0.2−0.4	0.96	0.95	0.86	0.98	0.90

[*]Clinical evaluation (at TMP = 50 mm Hg)

Table 9
Definition of Cryogel and Cryoprotein

Cryogel
    material gelled or aggregated
    from heparinized plasma by
    rapid cooling.

Cryoprotein
    protein precipitated
    from serum utilizing refrigeration
    for 3 to 5 days.

---

solution based upon molecular size and shape; precipitation is not required for cryogel to form. Cryogel is formed even in normal plasma. Cryogel is a mixture of various types of macromolecules including fibrinogen, globulin fractions, immune complexes, antibodies, and albumin. Cryogel formation is dynamic and is expected to be a function of operating parameters, membrane structure, and the composition of the plasma being treated.

CLINICAL TREATMENT OF RHEUMATOID ARTHRITIS

Plasmapheresis had not been used prior to 1979 for treatment of rheumatoid arthritis except for Jaffe's early work.[89] Since successful results in treating rheumatoid arthritis by therapeutic plasmapheresis [90,91] and lymphoplasmapheresis[92] were reported in 1979, many clinical treatments have been performed with recognizable improvements of patients.[93-95] These results suggest the applicability of plasmapheresis for the treatment of rheumatoid arthritis, while also indicating that the removal of pathologic macromolecules such as immune complexes, rheumatoid factors, and cryoglobulins plays a role in the clinical improvement seen. While these initial clinical trials of plasmapheresis were underway, the first clinical application of cryofiltration was performed in a 62-year-old female known to have rheumatoid arthritis since the age of 15.[21,96] This patient, who had severe unremitting disease with high levels of circulating immune complexes and cryoglobulins, was treated three times in the first week and was maintained on a twice weekly schedule for 10 weeks. Less frequent treatments have been applied during the last 2 years. Currently, she is on the treatment every 2 months and maintains a stable improved condition (Fig. 6). Six weeks after initiation of plasmapheresis with cryofiltration, circulating immune complex levels were reduced to 25% of the initial value (2,300 U/ml) and cryoprotein concentrations to 10% (initial value 3.0 mg/ml) using this technique alone. The rheumatoid factor remained elevated and showed only a slight decrease despite significant removal by the cryofilter in each treatment. Hematological studies did not indicate any significant changes in platelet count. White blood cell counts were observed in the early perfusions to decrease within the first 30 minutes of perfusion, but reached values higher than the preperfusion values by the end of the treatment.[97] Clinical measures of disease activity including morning stiffness and subjective state of pain improved significantly.

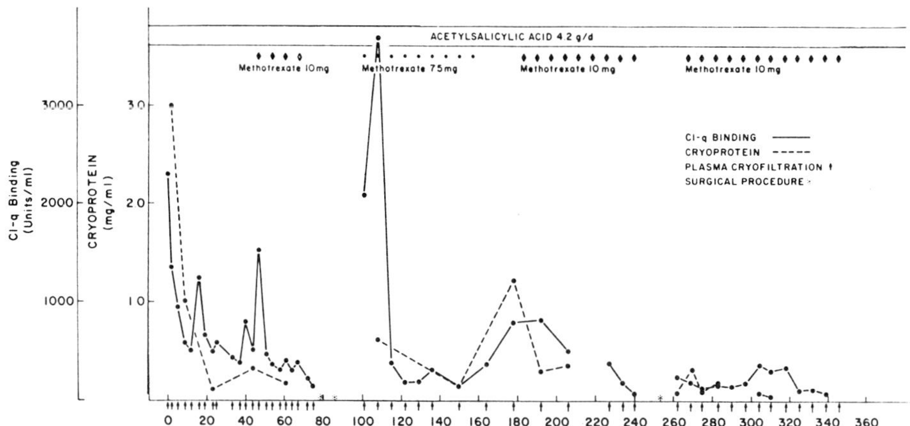

Figure 6: *Changes in pretreatment Clq-binding immune complex and cryoprotein levels for the first year of cryofiltration therapy in a patient with rheumatoid arthritis.*

Since the first clinical treatment of a patient with rheumatoid arthritis in April 1980, cryofiltration has been applied clinically in 13 patients with rheumatoid arthritis at the Cleveland Clinic through December 1982 (Table 10). Patients are selected who demonstrate seropositive rheumatoid arthritis as evaluated by elevated levels of rheumatoid factor, Clq-binding immune complexes, Westergren sedimentation rate, and cryoglobulins and who are failing or had failed at least one major antirheumatic drug therapy. Other criteria for inclusion in the study are that they have evaluable joint involvement (not class IV) with synovitis and x-ray demonstrated erosion. The protocol used for evaluation of the cryofiltration

Table 10
List of Patients Treated By Cryofiltration
at Cleveland Clinic (By Dec. 12, 1982)

Patient I.D.		Age	Sex	Disease	Start M/D/Y	Times
1	MW	62	F	RA	4/28/80	76
2	RE	65	M	RA(RV)	6/23/81	10
3	VG	55	F	RA	8/31/81	8
4	AZ	59	F	RA	10/30/81	11
5	MW	50	F	RA	1/11/82	14
6	DJ	59	F	RA	1/15/82	31
7	LT	27	M	RA	2/18/82	20
8	WE	50	F	RA	3/26/82	20
9	ST	41	F	RA	3/29/82	22
10	VM	42	M	RA	3/31/82	19
11	PM	36	F	RA	5/17/82	9
12	MW	57	F	RA	8/09/82	18
13	DG	48	F	RA	10/18/82	6
14	AT	34	M	SC	5/13/81	8
15	ES	85	M	CIHA	5/27/81	2
16	LW	54	M	CALD	7/23/81	1
17	MT	65	F	CG	3/08/82	3
18	VS	19	F	SLE	6/02/81	1
19	EH	59	F	CG	6/17/82	3
20	AW	44	F	SLE	6/22/82	4
21	PG	46	F	CG	11/09/82	2

RA = Rheumatoid arthritis; SC = Sclerosing cholangitis; CHIA = Cold-induced hemolytic anemia; CALD = Chronic alcoholic liver disease; CG = Cryoglobulinemia SLE = Systemic lupus erythematosus.

system consists of the treatment of patients twice per week for 5 weeks. Each procedure consists of the processing of 3 to 5 liters of plasma in a period of 3 to 4 hours.

Clinical evaluations were performed with six patients who received at least 10 treatments using the Asahi Hi-05 primary filter and Asahi Plasmaflo cryofilter.[98] Substances that were monitored during these evaluations include: albumin, total protein, immunoglobulins, Clq-binding immune complexes, rheumatoid factor, quantitative cryoglobulins, Westergren sedimentation rate, and hematology profiles. In addition to biochemical and hematologic assays, clinical evaluations were performed prior to the first treatment and every other treatment thereafter. Changes in biochemical parameters are summarized for the first and tenth treatments in Table 11.[98] Significant decreases in most values including rheumatoid factor, Clq-binding immune complexes, IgG, and IgM were seen. Concerning clinical parameters, Figure 7 illustrates changes that occurred in the Ritchie Index and the duration of morning stiffness.

Treatment of 15 patients with rheumatoid arthritis in a three-center study (6 patients at Cleveland Clinic, Cleveland; 5 patients at Cedar-Sinai Medical Center, Los Angeles; 4 patients at Rush Presbyterian St. Luke's Medical Center, Chicago) was begun.[99] Only seropositive patients with active disease, synovitis, and erosions were selected for this study. Twelve of the patients showed improvement of morning stiffness; in five of these patients, morning stiffness of several hours' duration nearly disappeared. All patients showed improved articular index usually in a regular stepwise fashion after each consecutive treatment. The average reduction of Clq binding during each treatment was 60%, and cryoglobulins present in significant concentrations in four patients were undetectable after five treatments. An additional seven patients (total 13 patients) have been treated at the Cleveland Clinic since February 1981 (Table 10).[100]

Results of hematologic studies were described for the three-center study[28,32] for long-term treatment at the Cleve-

Table 11
Immunochemical Parameters: Effects of Cryofiltratin in Rheumatoid Arthritis Patients[98]

Patient	Treatment Number	Immune Complexes (U/ml)	Rheumatoid Factor (RLS)	IgG (mg/dl)	IgM (mg/dl)	Albumin (mg/dl)	Fibrinogen (mg/dl)
1	1	2296	252	738	265	3.7	-
	10	441	237	474	154	3.1	410
2	1	104	213	769	110	3.3	362
	10	53	39	381	60	3.2	310
3	1	85	144	894	362	3.8	390
	10	73	117	859	284	3.8	430
4	1	135	175	648	231	4.4	285
	10	68	125	491	168	3.8	-
5	1	242	180	965	265	3.5	465
	10	56	77	576	146	2.9	303
6	1	58	46	729	136	3.5	525
	10	47	79	805	115	3.2	286
Mean ± SE	1	487 ± 889	168 ± 70	790 ± 117	228 ± 93	3.7 ± 0.4	405 ± 93
	10	123 ± 156	112 ± 68	598 ± 192	154 ± 74 $p < 0.05$	3.3 ± 0.4 $p < 0.05$	348 ± 67

(From Smith JW, Kayashima K, Katsume C, et al: Cryopheresis: Immunochemical modulation and clinical response in autoimmune disease. *Trans Am Soc Artif Organs* 1982; 28:391. By permission of the American Society for Artificial Internal Organs.)

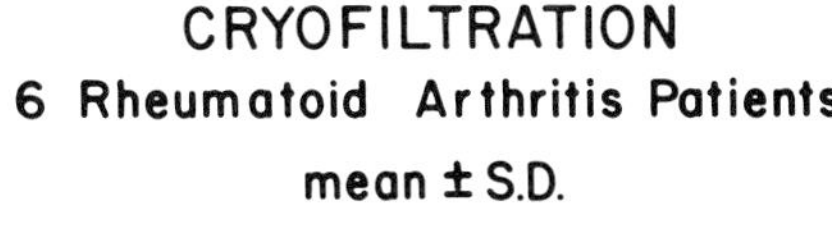

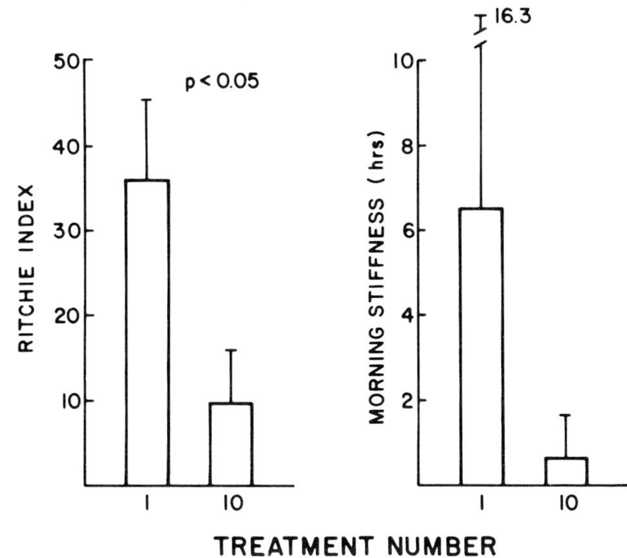

Figure 7: *Values of the Ritchie Index and length of morning stiffness in six rheumatoid arthritis patients at the first and tenth cryofiltration procedure.*

land Clinic[101] and in a special study concerning phagocytic function during cryofiltration.[102] The three-center study indicates that a modest decline in white blood cell counts occurred from $11,200 \pm 2,930/mm^3$ before treatment to $8,290 \pm 3,620/mm^3$ one week after the complete therapeutic regimen. Serial white blood cell measurements during a single procedure disclosed a decline in numbers of about 50% during the initial 30 minutes with return to pretreatment values after 50 minutes. This white blood cell decline was paralleled by a transient decrease of C3 to about 54% of pretreatment values. Changes in lymphocyte counts and platelet counts $(1,610 \pm 940/mm^3$ versus $1,260 \pm 750/mm^3$ lymphocytes; $407,000 \pm 216/mm^3$ versus $399,000 \pm 154/mm^3$ platelets) were not significant following treatment. In long-term evaluations (four rheumatoid patients treated by cryofiltration more than 20 times during a period of up to 30 months) of hematological changes, no significant changes in platelet and red blood cell counts occurred while a 15% decrease of white blood cell counts was reported.[101] It has been pointed out that early leukopenia accompanied by complement activation and secondary leukocytosis is related to phagocytic polymorphonuclear leukocyte activity. This may lead to additional beneficial effects by further lowering macromolecules such as immune complexes from plasma.[102] Recovery of T-helper and T-suppressor cell function after plasmapheresis was also reported in treating rheumatoid arthritis[95] and systemic lupus erythematosus[103] by plasma exchange. This suggests that cryofiltration may rapidly reverse reticuloendothelial blockage with resultant clinical benefit in the same manner as plasma exchange therapy.

Worldwide, more than 60 patients, most with rheumatoid arthritis, have been treated at five centers.[104-106] In Japan, 26 patients with rheumatoid arthritis have been treated more than 860 times by cryofiltration, averaging 4.6 liters of plasma filtration per session. These patients were evaluated according to the criteria described by Wallace et al.[92] as follows: excellent, 9 patients (34.6%); good, 4 patients (15.3%); fair, 7 patients (26.9%); poor, 5 patients (19.2%); undefined, 1 patient (3.8%). The Japanese investigators believe more than 25 cryofiltrations are required for effective clinical results.[105]

Overall results from these clinical treatments indicate that immunochemical parameters such as circulating immune complexes are reduced and quantitative cryoproteins are reduced to essentially normal levels.

OPERATIONAL AND DESIGN CONSIDERATONS OF CRYOFILTRATION

While the presently used cryofilter (Plasmaflo) was not originally designed for this application, good clinical results have been seen with its use. Generallly, early plugging of the filter occurs, necessitating module replacement during treatment. From extraction studies,[107] it has been shown than 20 to 25 g of protein are removed from rheumatoid arthritis patients per cryofilter used. A number of special studies of cryofilter operation and design have been carried out. Important considerations include: (1) temperature; (2) plasma constituents due to type of disease; (3) anticoagulants; (4) membrane structure (pore size, porosity, depth of membrane); and (5) flow conditions including flow systems.

Temperature

Table 12 shows changes of transmembrane pressure in plasma filtration with normal human plasma at varying temperatures.[108] As the temperature is reduced, the volume filtered is reduced and the removal per volume processed is increased. Table 13 outlines the volume processed to reach a pressure of 300 mm Hg mean sieving for solutes and total removal rates at varying temperatures using normal human plasma. Even at 35°C, substantial amounts of plasma solutes are removed. Sieving properties of total protein from normal plasma is generally in the range of 0.8 below 25°C. At 37°C, the sieving coefficient of total protein is in the range of 0.95. Thus, even without active cooling of the plasma (i.e., the cryofilter is operated at ambient temperature), cryogel is still produced.

Table 12
Filtration Volume Versus Transmembrane Pressure
Normal Human Plasma[87]

Temp (°C)	Filter	100 mm Hg	200 mm Hg	300 mm Hg
-1	Asahi Plasmaflo	495	574	755
4	"	923	1154	1475
25	"	1375	1775	2020
37	"	—	—	—

Table 13
Differences in Volume Processed, Mean Sieving Coefficients, and Total Removal
for Normal Plasma Treated by Cryofiltration at Varying Temperatures

Temperature, °C	Volume processed[+] ml/module	ml/m^2	Mean Sieving Coefficient			Total Removal (g/module, liter)		
			Total protein	Albumin	Total globulin	Total protein	Fibrinogen	Total cholesterol
-1	755	1162	0.82	0.87	0.72	8.00	1.12	0.43
4	1475	2269	0.82	0.85	0.77	10.29	0.44	0.50
25	2020	3108	0.78	0.81	0.72	11.73	0.97	0.61
37*	2060	3169	0.95	0.96	0.91	2.75	2.43	0.41

[+]to pressure of 300 mm Hg;
*to pressure of 34 mm Hg

Operation at ambient temperature can lead to variable performance as reported by several investigators.[108,109] It is advisable to cool the plasma and cryofilter to below 25°C for effectiveness. The recommended temperature range is from 0° to 10°C. During clinical operation where the sieving of macromolecules cannot be measured, monitoring the transmembrane pressure of the cryofilter provides a useful guide of the clinical efficiency.

Effect of Disease

Cryogel formation and solute removal is higher if pathologic plasmas are used.[110,111] Figure 8 outlines differences in the filtrate volume for the Plasmaflo studied with various disease state plasmas. Cryogel removal is higher in rheumatoid arthritis, Sjögren's syndrome, and sclerosing cholangitis compared to normal human plasma. Sieving properties also

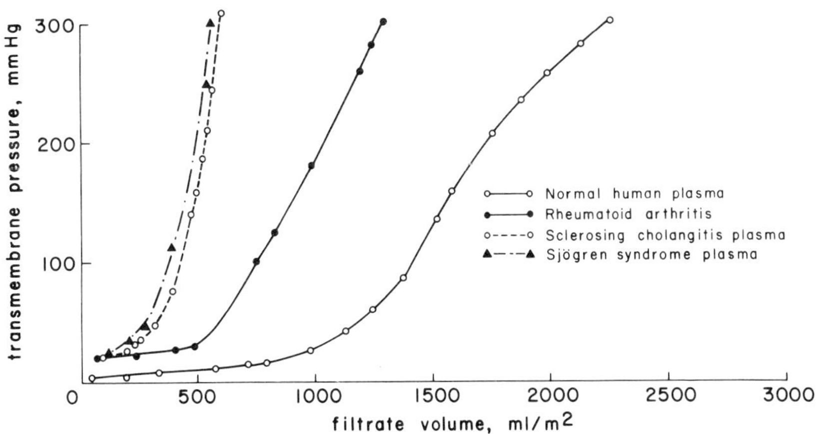

Figure 8: *Changes in transmembrane pressure with normalized filtrate volume for plasma from patients with various diseases filtered through the Plasmaflo cryofilter at 4°C. (From Nosé Y, et al: Therapeutic cryogel removal in autoimmune disease: What is cryogel? In* Therapeutic Plasmapheresis (II). *Edited by T. Oda. Stuttgart, FK Schattauer Verlag, 1982. By permission of publisher.)*

indicate specific removal due to type of disease states as shown in Figure 9. Albumin macromolecule separation efficiency (SE) defined as SE = [(albumin sieving - globulin sieving)/albumin sieving)] × 100, for globulin was 49% in Sjögren's syndrome, 29% in rheumatoid arthritis, 25% in sclerosing cholangitis, and 13% in normal human plasma below 25°C and 5% in normal human plasma at 37°C. SE for fibrinogen was 43% in Sjögren's syndrome, 57% in rheumatoid arthritis, 78% in sclerosing cholangitis, and 43% in normal human plasma at 37°C. In plasma from a patient with Sjögren's syndrome and cryoglobulinemia, the main constituents removed by cryofiltration are immunoglobulins; cholesterol and fibrinogen dominate cryogel formation in sclerosing cholangitis. On the other hand, fibrinogen and globulin fractions are removed preferentially over albumin in rheumatoid arthritis. The abnormal

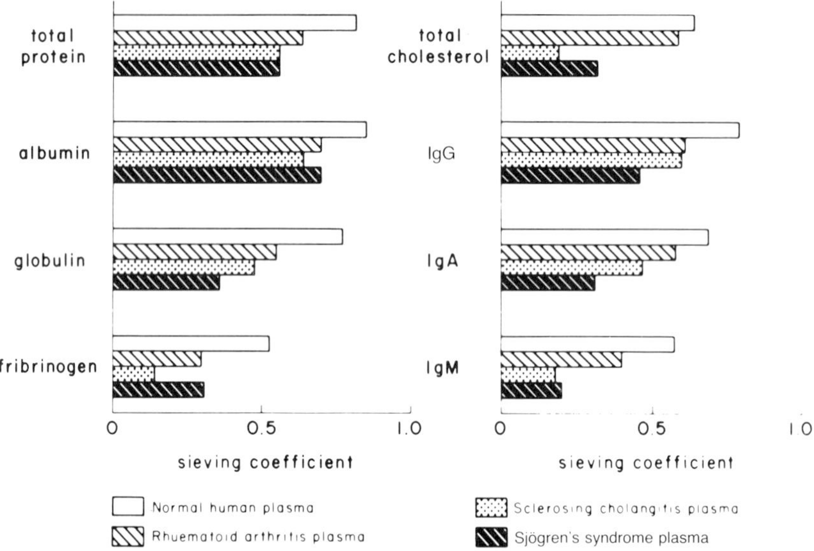

Figure 9: *Mean sieving coefficients of various solutes for disease state plasmas and normal human plasma filtered through the Plasmaflo cryofilter at 4°C. (From Nosé Y, et al: Therapeutic cryogel removal in autoimmune disease: What is cryogel? In* Therapeutic Plasmapheresis (II). *Edited by T Oda. Stuttgart, FK Schattauer Verlag, 1982. By permission of publisher.)*

solubility of immunoglobulins in cryoglobulin-rich Sjögren's syndrome plasma and the association of lipids with fibrinogen in sclerosing cholangitis are considered as hypotheses of cryogel formation in these cases.[112] The mechanisms of cryogel formation in rheumatoid arthritis are still unclear; however, fibrinogen seems to contribute to cryogel formation, because significant removal of fibrinogen occurs not only during in vitro cryofiltration (30% sieving) but also during clinical studies (47% reduction after treatment). One mechanism that is well documented for cold-induced precipitation of macromolecules from plasma is the interaction of heparin, fibronectin (cold insoluble globulin), and fibrin/ fibrinogen under cooling conditions to form a precipitable complex.[113,114] It is hypothesized that this complex acts as a nucleus for cryogel formation and the binding of macromolecular weight plasma solutes which are then retained in the cryofilter.

Effect of Anticoagulant

As mentioned in the preceding section, the concentration of heparin may be an important factor for cryofiltration. For extracorporeal circulation, the patient's blood is heparinized. Plasma subjected to the cooling procedure contains sufficient amounts of heparin to produce a heparin-fibronectin complex. The literature indicates that in the presence of a certain level of heparin, fibrinogen aggregates with fibronectin and precipitates from plasma at cold temperature.[115,116] With a heparin level of 0.05 mg/ml and a temperature of 2° to 4°C, this complex is formed within 20 minutes, comparable to conditions used in our clinical procedures. It is also known from in vitro studies that immune complexes are involved in cryoprecipitable substances.[117] Trapping of immune complexes and fibrinogen is good, as evidence by the low sieving coefficient in the cryofilter and high percent reductions for the entire procedure. It should be noted that cryogel formation takes place even using normal heparinized plasma and is effected by heparin concentration[118] as indicated in Table 14.

Effect of Membrane Properties

While plasma types and temperature are important[108-111] to cryogel formation, membrane pore size and microstructure are important to cryogel retention. From the results of clinical trials and in vitro filtration studies, two cryofilters (Asahi Plasmaflo) are typically required to filter 3 to 5 liters of plasma in treating rheumatoid arthritis patients, and several filters are required to treat patients with Sjögren's syndrome with high levels of cryoglobulin or to treat patients with sclerosing cholangitis and high lipoprotein levels. To avoid changing the cryofilter during treatment, large pore size membranes and prefilters are under development. In preliminary cryofiltration tests, membranes with one micron pore size can process large volumes of plasma, with good removal of Clq-binding immune complexes and cryoglobulins from plasma from a patient with Sjögren's syndrome. The combination of a pre-filter (pore size greater than 0.4 μm) with the cryofilter also allows greater plasma processing by the cryofilter.

Table 14
Effects of Heparin Dose on Cryofiltration at 4°C

Heparin Dose (units/liter)	Membrane Pore Size (μm)	Temperature (°C)	Normalized Volume Processed (ml/m²)	Type of Plasma
2000	0.2	4	285	Fresh Bovine Plasma
5000	0.2	4	2250	Fresh Bovine Plasma
10000	0.2	4	8300	Fresh Bovine Plasma
2000	0.2	4	18500	Fresh Bovine Serum

Flow Conditions

Flow systems through the membrane are classified into single-pass method, single-pass with partial discard method, and recirculating method. In general, single-pass with partial discard allows greater volume processing but produces significant amounts of protein losses. The recirculating method is effective in preventing concentration polarization on the membrane surface[119] but requires at least two pumps. Each of these flow methods has been used clinically, but single-pass is still preferred due to simplicity and reduced protein loss. Selection of flow systems may be less important clinically than other factors mentioned above.

CONCLUSION

Plasma exchange has been employed in the treatment of various diseases for the removal of circulating immune complexes, antibodies, inflammatory mediators, and other pathologic macromolecular solutes. The literature cites over 50 diseases treated by plasmapheresis. However, the pathogenesis of many of the diseases being treated is not clear, and the relationship of so-called pathologic macromolecules that are used as treatment indicators to the disease state is not well understood.

An on-line cryofiltration system was originally used in a patient with rheumatoid arthritis who had high cryoglobulin and circulating immune complex levels. Successful results obtained with this first clinical trial[21] encouraged the extended application of this therapy. Since 1980 more than 60 patients have been treated by cryofiltration systems at five institutions. Overall results of the clinical study indicate that significant improvements of morning stiffness and Ritchie Index occur, accompanied by reduction of Clq-binding immune complexes, cryoglobulins, and other circulating plasma factors that may mediate disease activity. This on-line removal technique involves the formation of cryogel from the heparinized plasma, with trapping of the cryogel on the macromolecule filter (cryofilter). This process eliminates or minimizes requirements for replacement plasma products.

Although a relatively new technique, clinical improvement has been seen in patient studies to date. While analogous to the early days of hemodialysis in terms of technical capabilities and use of presumed pathogenic markers, it is our hope that investigation and use of techniques such as cryofiltration will aid in the elucidation of disease etiology and appropriate markers to follow disease activity, providing a sound, scientific basis for the use of apheresis techniques.

REFERENCES

1. Horiuchi T, Kambic H, Takatani S, et al: *Topics in Plasmapheresis: A bibliography of Therapeutic Applications and New Techniques*. International Center for Artificial Organs and Transplantation, Cleveland, 1982.
2. Kambic H, Nosé Y: *Plasmapheresis: Historical Perspective, Therapeutic Applications and New Frontiers*. International Center for Artificial Organs and Transplantation, Cleveland, 1982.
3. Schwab PJ, Fahey JL: Treatment of Waldenström's macroglobulinemia by plasmapheresis. *N Engl J Med* 1960; 263:574.
4. Skoog WA, Adams WS, Coburn JW: Metabolic balance study of plasmapheresis in a case of Waldenström's macroglobulinemia. *Blood* 1962; 19:425.
5. Solomon A, Fahey JL: Plasmapheresis therapy in macroglobulinemia. *Ann Intern Med* 1963; 58:789.
6. Friedman BA, Schork MA, Mocniak JL, et al: Short-term and long-term effects of plasmapheresis on serum proteins and immunoglobulins. *Transfusion* 1975; 15:467.
7. Keller AJ, Chirnside A, Urbaniak SJ: Coagulation abnormalities produced by plasma exchange on the cell separator with special reference to fibrinogen and platelet levels. *Br J Haematol* 1979; 42:593.
8. Ashkar FS, Katims RB, Smoak WM III, et al: Thyroid storm treatment with blood exchange and plasmapheresis. *JAMA* 1970; 214:1275.
9. Branda RF, Moldow CF, McCullough JJ, et al: Plasma exchange in the treatment of immune disease. *Transfusion* 1975; 15:570.
10. Cardella CJ, Sutton D, Uldall PR, et al: Intensive plasma exchange and renal-transplant rejection. *Lancet* 1977; 1:264.

11. Israel L, Edelstein R, Mannoni P, et al: Plasmapheresis and immunological control of cancer. *Lancet* 1976; 2:642 (letter to the editor).
12. Jones JV, Cumming RH, Bucknall RC, et al: Plasmapheresis in the management of acute systemic lupus erythematosus? *Lancet* 1979; 1:709.
13. Lepore MJ, Martel AJ: Plasmapheresis with plasma exchange in hepatic coma: Methods and results in five patients with acute fulminant hepatic necrosis. *Ann Intern Med* 1970; 72:165.
14. Lockwood CM, Boulton-Jones JM, Lowenthal RM, et al: Recovery from Goodpasture's syndrome after immunosuppressive treatment and plasmapheresis. *Br Med J* 1975; 2:252.
15. Pinching AJ, Peters DK, Davis JN: Remission of myasthenia gravis following plasma-exchange. *Lancet* 1976; 2:1373.
16. Ruocco V, Rossi A, Argenziano G, et al: Pathogenicity of the intercellular antibodies of pemphigus and their periodic removal from the circulation by plasmapheresis. *Br J Dermatol* 1978; 98:237.
17. Gurland H, Samtleben W, Blumenstein M: Review article: Therapeutic plasmapheresis; present state and future aspects. *Life Support Systems* 1983; 1:61.
18. Jones JV, McLeod BC: Development and assessment of systems for removal of plasma components. *Artif Organs* 1981; Suppl 5:117.
19. Pineda AA, Taswell HF: Selective plasma component removal: Alternatives to plasma exchange. *Artif Organs* 1981; 5:234.
20. Malchesky PS, Asanuma Y, Zawicki I, et al: On-line separation of macromolecules by membrane filtration with cryogelation. *Artif Organs* 1980; 4:205.
21. Malchesky PS, Asanuma Y, Zawicki I, et al: On-line separation of macromolecules by membrane filtration with cryogelation. **In** *Plasma Exchange: Plasmapheresis, Plasma Separation.* Edited by HG Sieberth. Stuttgart, FK Schattauer Verlag, 1980, p 133.
22. Nosé Y, Malchesky PS, Asanuma Y: Augmented solute reduction in diseases treated by extracorporeal detoxification systems: X-effect hypothesis. **In** *Artificial Liver Support.* Edited by G Brunner, FW Schmidt. Berlin, Springer-Verlag, 1981, p 181.
23. Nosé Y, Malchesky PS, Smith JW: Hybrid artificial organs: Are they really necessary? *Artif Organs* 1980; 4:285.
24. Handley SL, Vogel RA: *Therapeutic apheresis: Industry Research.* Unterberg, Towbin, LF Rothschild, September 1981.
25. Scoville PL: *Therapeutic Plasmapheresis: Review and Market Forecast.* Greenville, SC, Phillips L Scoville Associates, February 1981.
26. Yatzidis HA: A convenient haemoperfusion micro-apparatus over charcoal for the treatment of endogenous and exogenous intoxication; Its use as an effective artificial kidney. *Proc Eur Dial Transplant Assoc* 1964; 1:83.
27. Asanuma Y, Malchesky PS, Smith JW, et al: Removal of protein-bound toxins from critical care patients. *Clin Toxicol* 1980; 17:571.
28. Asanuma Y, Malchesky P, Smith J, et al: Chronic ambulatory liver support by membrane plasmapheresis with on-line detoxification. *Trans Am Soc Artif Intern Organs* 1981; 27:416.

29. Asanuma Y, Smith JW, Malchesky PS, et al: Preclinical evaluation of membrane plasmapheresis with on-line bilirubin removal. *Artif Organs* 1979: Suppl 3:279.

30. Asanuma Y, Smith JW, Suwa S, et al: Membrane plasmapheresis: Platelet and protein effects on filtration. *Proc Eur Soc Artif Organs* 1979; 6:308.

31. Asanuma Y, Malchesky PS, Zawicki I, et al: Clinical hepatic support by on-line plasma treatment with multiple sorbents: Evaluation of system performance. *Trans Am Soc Artif Intern Organs* 1980; 26:400.

32. Carey WD, Nosé Y, Ferguson DR, et al: Plasma perfusion in liver disease: Phase I study. **In** *Plasma Exchange: Plasmapheresis, Plasmaseparation*. Edited by HG Sieberth. Stuttgart, FK Schattauer Verlag, 1980, p 335.

33. Carey WD, Smith J, Asanuma Y, et al: Pruritus of cholestasis treated with plasma perfusion. *Am J Gastroenterol* 1981; 76:330.

34. Castino F, Scheucher K, Malchesky PS, et al: Microemboli-free blood detoxification utilizing plasma filtration. *Trans Am Soc Artif Intern Organs* 1976; 22:637.

35. Malchesky PS, Asanuma Y, Hammerschmidt DE, et al: Complement removal of sorbents in membrane plasmapheresis with on-line plasma treatment. *Trans Am Soc Artif Intern Organs* 1980; 26:541.

36. Malchesky PS, Asanuma Y, Smith J, et al: Membrane plasmapheresis with on-line plasma treatment. **In** *Hemoperfusion: Kidney and Liver Support and Detoxification*, Part 1. Edited by S Sideman, TMS Chang. Washington, DC, Hemisphere Publishing Corporation, 1980, p 111.

37. Malchesky PS, Ouchi K, Piatkiewicz W, et al.: Membrane plasma filtration systems with multiple reactors for hepatic support. *Artif Organs* 1978; Suppl 2:265.

38. Nosé Y, Koshino I, Castino F, et al: Further assessment of liver tissue materials for extracorporeal hepatic assist. **In** *Artificial Liver Support*. Edited by R Williams, IM Murray-Lyon. London, Pitman Medical Press, 1975, p 202.

39. Nosé Y, Malchesky PS, Asanuma Y, et al: Procedures and methodology of hemoperfusion as hepatic assist. **In** *Hemoperfusion: Kidney and Liver Support and Detoxification*. Part 1. Edited by S Sideman, TMS Chang. Washington, DC, Hemisphere Publishing Corporation, 1980, p 265.

40. Nosé Y, Malchesky PS, Asanuma Y, et al: Plasma filtration detoxification on hepatic patients: Its optimal operating conditions. **In** *Therapeutic Plasma Exchange*. Edited by HJ Gurland, V Heinze, HA Lee. Berlin, Springer-Verlag, 1981, p 125.

41. Ouchi K, Malchesky PS, Piatkiewicz W, et al: Improved hemoperfusion sorbents and plasma filtration system for hepatic assist. *Proc Am Con Engl Med Biol* 1977; 19:439.

42. Smith JW, Asanuma Y, Malchesky PS, et al: Treatment of hepatic dysfunction using membrane plasmapheresis with sorptive plasma detoxification. *Artif Organs* 1981; Suppl 5:828.

43. Smith JW, Matsubara S, Horiuchi T, et al: Sorption-filtration therapy

for chronic liver disease: In vitro testing and clinical correlation. *Trans Am Soc Artif Intern Organs* 1982; 28:215.

44. Bansal SC, Bansal BR, Thomas HL, et al: *Ex vivo* removal of serum IgG in a patient with colon carcinoma: Some biochemical, immunological and histological observations. *Cancer* 1978; 42:1.

45. Bensinger WI, Baker DA, Buckner CD, et al: Immunoadsorption for removal of A and B blood-group antibodies. *N Engl J Med* 1981; 304:160.

46. Borberg H, Grewe V, Sawatzki C, et al: Specific continuous flow immunoadsorption ex vivo in patients with familial hypercholesterolaemia. **In** *New Membranes in Medical Treatment*. Second Tutzing Symposium on Chemical Engineering in Medicine. 1981, p 58 (abstract).

47. Burgstaler EA, Pineda AA, Ellefson RD: Removal of plasma lipoproteins from circulating blood with a heparin-agarose column. *Mayo Clin Proc* 1980; 55:180.

48. Chang TMS: Blood compatible coating of synthetic immunoadsorbents. *Trans Am Soc Artif Intern Organs* 1980; 26:546.

49. Ray PK, Idiculla A, Rhoads JE Jr, et al: Immunoadsorption of IgG molecules from the plasma of multiple myeloma and autoimmune hemolytic anemia patients. *Plasma Ther* Feb. 1980; 1:11.

50. Rawer P, Sommerlad KH, Goretzki K, et al: Elimination of plasma proteins during plasma separation using either centrifugation or membrane filtration and selective removal of immunoglobulin G and immune complexes with immobilized protein A. **In** *Plasma Exchange: Plasmapheresis, Plasmaseparation*. Edited by HG Sieberth. Stuttgart, FK Schattauer Verlag, 1980, p 65.

51. Schmer G, Newman ML, Rastelli L, et al: Glutaraldehyde-treated xenocells: A specific absorbent for complement factors. *Trans Am Soc Artif Intern Organs* 1981; 27:445.

52. Schmer G, Rastelli L, Dennis MB Jr, et al: Removal of low density lipoproteins in sheep by extracorporeal affinity chromatography. *Trans Am Soc Artif Intern Organs* 1981; 27:547.

53. Stoffel W, Demant TH, Sieberth HG, et al: Selective removal of Apo B containing serum lipoproteins from blood plasma. **In** *Plasma Exchange: Plasmapheresis, Plasmaseparation*. Edited by HG Sieberth. Stuttgart, FK Schattauer Verlag, 1980, p 127.

54. Terman DS: Chemotherapeutic agents for specific suppression of antibody rebound after extracorporeal immunoadsorption (EI). *Muscle Nerve* 1978; 1:341 (abstract).

55. Terman DS, Buffaloe G, Mattioli C, et al: Extracorporeal immunoadsorption: Initial experience in human systemic lupus erythematosus. *Lancet* 1979; 2:824.

56. Terman DS, Durante D, Buffaloe G, et al: Attenuation of canine nephrotoxic glomerulonephritis with an extracorporeal immunoadsorbent. *Scand J Immunol* 1977; 6:195.

57. Terman DS, Yamamoto T, Tillquist RL, et al: Tumoricidal response induced by cytosine arabinoside after plasma perfusion over protein A. *Science* 1980; 209:1257.

58. Terman DS, Young JB, Shearer WT, et al: Preliminary observations of the effects on breast adenocarcinoma of plasma perfused over immobilized protein A. *N Engl J Med* 1981; 305:1195.

59. Nosé Y, Gurland HJ, Klein HG, et al: Plasmapheresis and cytapheresis. *Trans Am Soc Artif Intern Organs* 1982; 28:631.

60. Zubler RH, Lambert PH: Detection of immune complexes in human diseases. *Prog Allergy* 1978; 24:1.

61. Cornwell DG, Kruger FA: Molecular complexes in the isolation and characterization of plasma lipoproteins. *J Lipid Res* 1961; 2:110.

62. Malchesky PS, Asanuma Y, Smith JW, et al: Macromolecule removal from blood. *Trans Am Soc Artif Intern Organs* 1981; 27:439.

63. Agishi T, Kaneko I, Hasuo Y, et al: Double filtration plasmapheresis. *Trans Am Soc Artif Intern Organs* 1980; 26:406.

64. Fassbinder W, Platzer E, Ernst W, et al: Immune complex elimination by plasma separation with membrane filtration. **In** *Plasma Exchange: Plasmapheresis, Plasmaseparation*. Edited by HG Sieberth. Stuttgart, FK Schattauer Verlag, 1980, p 107.

65. Glöckner WM, Sieberth HG, Dienst C, et al: Elimination kinetics of antibodies and immune complexes in membrane plasma separation. **In** *Plasma Exchange: Plasmapheresis, Plasmaseparation*. Edited by HG Sieberth. Stuttgart, FK Schattauer Verlag, 1980, p 121.

66. Sieberth HG: *Plasma Exchange: Plasmapheresis, Plasmaseparation*. Stuttgart, FK Schattauer Verlag, 1980, p 29.

67. Sieberth HG, Glöckner WM, Dienst C, et al: Cascade filtration in man. *Artif Organs* 1981; Suppl 5:122.

68. Euler HH, Béress R, Münster F, et al: Experimental elimination characteristics of high molecular substances and radiolabelled immune complexes in hollow fiber plasmapheresis filters. *Artif Organs* 1981; Suppl 5:92.

69. Kayashima K, Sueoka A, Smith JW, et al: Development of new hollow fiber membrane macromolecular filters. *Trans Am Soc Artif Intern Organs* 1982; 28:66.

70. Cream JJ: Cryoglobulins in vasculitis. *Clin Exp Immunol* 1972; 10:117.

71. Brouet J-C, Clauvel J-P, Danon F, et al: Biologic and clinical significance of cryoglobulins: A report of 86 cases. *Am J Med* 1974; 57:775.

72. McIntosh RM, Koss MN, Gocke DJ: The nature and incidence of cryoproteins in hepatitis B antigen (Hb$_s$Ag) positive patients. *Q J Med* 1976; 45:23.

73. Erhardt CC, Mumford P, Maini RN: The association of cryoglobulinaemia with nodules, vasculitis and fibrosing alveolitis in rheumatoid arthritis and their relationship to serum Clq binding activity and rheumatoid factor. *Clin Exp Immunol* 1979; 38:405.

74. Weisman M, Zvaifler N: Cryoimmunoglobulinemia in rheumatoid arthritis: Significance in serum of patients with rheumatoid vasculitis. *J Clin Invest* 1975; 56:725.

75. McLeod BC, Sassetti RJ: Plasmapheresis with return of cryoglobulin-depleted autologous plasma (cryoglobulinpheresis) in cryoglobulinemia. *Blood* 1980; 55:866.

76. L'Abbate A, Paciucci A, Bartolomeo F, et al: Selective removal of plasma cryoglobulins in cryoglobulinaemia. *Proc Eur Dial Transplant Assoc* 1977; 14:486.

77. Maggiore Q, L'Abbate A, Caccamo A, et al: Cryopheresis in the treatment of essential mixed cryoglobulinemia with glomerulonephritis. *Artif Organs* 1981; Suppl 5:126.

78. Buchholz DH, Bove JR, Charette JR: Plasmapheresis using the IBM 2991 blood cell processor. *Transfusion* 1978; 18:269.

79. Holderman C, Schlesinger RG: Modified plasma therapy using the Haemonetics 30 blood processor. *Plasma Ther* 1981; 2:31.

80. Wojcicki JM, Malchesky PS, Sueoka A, et al: Analysis of filtration sensitivity in MPS. *Proc Ann Conf Engl Med Biol* 1982; 24:240.

81. Nosé Y, Kambic H, Matsubara S: Introduction to therapeutic apheresis. **In** *Plasmapheresis: Therapeutic Applications and New Techniques.* Edited by Y Nosé, PS Malchesky, JW Smith, et al. New York, Raven Press, 1983, p 1.

82. Malchesky PS, Werynski A, Asanuma Y, et al: Clinical operation of Asahi plasma separators. *Artif Organs* 1981; Suppl 5:113.

83. Malchesky PS, Nosé Y: Membranes for plasma separation and its treatment. **In** *World Filtration Congress III.* Vol. II. Croydon, England, Uplands Press, 1982, p 649.

84. Werynski A, Malchesky PS, Sueoka A, et al: Membrane plasma separation: Toward improved clinical operation. *Trans Am Soc Artif Intern Organs* 1981; 27:539.

85. Zawicki I, Malchesky PS, Smith JW, et al: Axial changes of blood and plasma flow, pressure, and cellular deposition in capillary plasma filters. *Artif Organs* 1981; 5:241.

86. Nosé Y, Malchesky PS: Technical aspects of membrane plasma treatment. *Artif Organs* 1981; Suppl 5:86.

87. Smith JW, Wysenbeek AJ, Krakauer RS: Plasmapheresis I: Membrane filtration methods. *Plasma Ther* July 1981; 2:53.

88. Nosé Y, Horiuchi T, Malchesky PS, et al: Therapeutic cryogel removal in autoimmune disease: What is cryogel? **In** *Therapeutic Plasmapheresis (II).* Edited by T Oda. Stuttgart, FK Schattauer Verlag, 1982, p 15.

89. Jaffe IA: Comparison of the effect of plasmapheresis and penicillamine on the level of circulating rheumatoid factor. *Ann Rheum Dis* 1963; 22:71.

90. Goldman JA, Casey HL, McIlwain H, et al: Limited plasmapheresis in rheumatoid arthritis with vasculitis. *Arthritis Rheum* 1979; 22:1146.

91. Rothwell RS, Davis P, Gordon PA, et al: A controlled study of plasma exchange in the treatment of severe rheumatoid arthritis. *Arthritis Rheum* 1980; 23:785.

92. Wallace DJ, Goldfinger D, Gatti R, et al: Plasmapheresis and lymphoplasmapheresis in the management of rheumatoid arthritis. *Arthritis Rheum* 1979; 22:703.

93. Jones JV, Clough JD, Klinenberg JR, et al: The role of therapeutic plasmapheresis in the rheumatic diseases. *J Lab Clin Med* 1981; 97:589.

94. Wallace DJ, Goldfinger D, Thompson-Breton R, et al: Advances in the use of therapeutic pheresis for the management of rheumatic disease. *Semin Arthritis Rheum* 1980; 10:81.

95. Yamagata J, Shiozawa K, Shiokawa Y: Therapeutic plasma exchange for rheumatic diseases. **In** *Plasma Exchange: Plasmapheresis, Plasmaseparation*. Edited by HG Sieberth. Stuttgart, FK Schattauer Verlag, 1980, p 265.

96. Krakauer RS, Asanuma Y, Zawicki I, et al: Circulating immune complexes in rheumatoid arthritis: Selective removal by cryogelation with membrane filtration. *Arch Intern Med* 1982; 142:395.

97. Asanuma Y, Malchesky PS, Blumenstein M, et al: Continuous cryofiltration (CCF) of plasma for rheumatoid arthritis. *Jpn J Artif Organs* 1981; 10:110.

98. Smith JW, Kayashima K, Katsume C, et al: Cryopheresis: Immunochemical modulation and clinical response in autoimmune disease. *Trans Am Soc Artif Organs* 1982; 28:391.

99. Krakauer RS, Wysenbeek AJ, Wallace DJ, et al: Therapeutic trial of cryofiltration in patients with rheumatoid arthritis. *Am J Med* 1983; 74:951.

100. Abe Y, Katsume C, Matsubara S, et al: Selective removal of immune complexes (IC) by cryofiltration in rheumatoid arthritis (RA). *ASAIO* 1983; 23:39 (abstract).

101. Ueno M, Smith JW, Matsubara S, et al: Hematological changes in long-term cryofiltration. *ASAIO* 1983; 12:86 (abstract).

102. Kayashima K, Smith JW, Barna R, et al: Improved white blood cell functions: Additional effects of membrane plasmapheresis. *Trans Am Soc Artif Intern Organs* 1982; 28:312.

103. Lockwood CM, Worlledge S, Nicholas A, et al: Reversal of impaired splenic function in patients with nephritis or vasculitis (or both) by plasma exchange. *N Engl J Med* 1979; 300:524.

104. Azuma N, Nobuto T, Suzuki M, et al: Clinical effect of plasmapheresis with continuous cryofiltration for three cases of rheumatoid arthritis. **In** *Therapeutic Plasmapheresis* (1). Edited by T Oda. Stuttgart, FK Schattauer Verlag, 1981, p 159.

105. Azuma N, Nobuto T, Suzuki M, et al: Clinical effects of continuous cryofiltration (CCF) on rheumatoid arthritis. *Jpn J Artif Organs* 1983; 12:287.

106. Shimo K, Suzuta T, Suzuki M, et al: Basic and clinical studies on continuous cryofiltration: A new treatment for rheumatoid arthritis. *Ryumachi* 1981; Suppl 21:61.

107. Katsume C, Horiuchi T, Kayashima K, et al: Quantification of cryogel in macromolecular filters. *ASAIO* 1982; 11:52 (abstract).

108. Horiuchi T, Malchesky PS, Sueoka A, et al: Analysis of cryogel for cryofiltration. *Jpn J Artif Organs* 1982; 11:1275 (abstract).

109. Sonoda T, Tanaka Y, Tanzawa H, et al: In vitro evaluation of PMMA macromolecular filter. *Jpn J Artif Organs* 1983; 12:121 (abstract).

110. Horiuchi T, Malchesky PS, Ueno M, et al: Quantitative analysis for on-line cryofiltration. *Proc Ann Conf Engl Med Biol* 1982; 24:232.

111. Horiuchi T, Ueno M, Katsume C, et al: Quantitative analysis of cryofiltration (CF) for the removal of pathologic substances. *ASAIO* 1982; 11:50 (abstract).

112. Middaugh CR, Lawson EQ, Litman GW, et al: Thermodynamic basis for the abnormal solubility of monoclonal cryoimmunoglobulins. *J Biol Chem* 1980; 255:6532.

113. Mosesson MW: Structure of human plasma cold-insoluble globulin and the mechanism of its precipitation in the cold with heparin or fibrin-fibrinogen complexes. *Ann NY Acad Sci* 1978; 312:11.

114. Mosesson MW, Amrani DL: The potential of heparin as an agent for precipitation of plasma fibronectin (Clq) and certain components of the plasma factor VIII complex[+]. **In** *Chemistry and Biology of Heparin.* Edited by RL Lundblad, WV Brown, KG Mann, et al. New York, Elsevier/North Holland, 1981, p 105.

115. Mosesson MW, Amrani DL: The structure and biologic activities of plasma fibronectin. *Blood* 1980; 56:145.

116. Smith RT, von Korff RW: A heparin-precipitable fraction of human plasma. I. Isolation and characterization of the fraction. *J Clin Invest* 1957; 36:596.

117. Hardin JA: Cryoprecipitogogue from normal serum: Mechanism for cryoprecipitation of immune complexes. *Proc Natl Acad Sci USA* 1981; 78:4562.

118. Katsume C, Abe Y, Matsubara S, et al: Cryogel studies for the optimization of cryofilration (CF) therapy. *ASAIO* 1983; 12:83 (abstract).

119. Porter MC: What, when and why of membranes MF, UF, and RO. **In** *What the Filterman Needs to Know about Filtration.* AIChE Symposium Series. 1977; 73:83.

Index